Korean Traditional Local Cuisines

한국전통 향토음식

국립농업과학원 지음

21세기사

Remarks on the Publication

Hansik (Korean food), which is mostly composed of fresh or fermented vegetables, is inherently good for one's health. It is not only pleasing on the eye with its diverse colors, it is also close to nature because it is cooked so as to bring out the inherent taste of the ingredients used, and presents a balanced and fresh diet. For the Koreans, food is equal to medicine. For these reasons, Hansik is regarded a desirable and ideal well-being food for the future.

As the Korean people have used crops native to their local area as their primary food source, the different provinces – the mountainous northeastern region, the coastal areas and islets of the western and eastern shores, the southwestern region that has vast plains – have all developed different varieties of foods and recipes.

In this book, we have carefully chosen 100 local foods representing the nine provinces of Korea. Every recipe, which is categorized by its native province, is introduced along with the ingredients, cooking directions, tips, and photos, so that readers can make Korean traditional foods by themselves much more conveniently.

We hope you can learn a great deal about the nature, culture and beauty of Korea and develop affection for Korea through the reading of this book.

Korean Food & Culture Division

National Academy of Agricultural Science
Rural Development Administration, Korea

Contents

1

Main Dishes on the Meal Table

Middle categories	Definition
1. Bab (Rice Dishes)	A gelatinized food made by pouring water of 1.2~1.5 times the amount of rice onto the rice and other grains. Sometimes dishes are made from vegetables, seafood or meat put together with the rice.
2. Juk (Porridge, Mush)	A nutritious formula made by pouring water of 6~7 times the amount of rice onto rice, barley, millet and so on and then boiling this for a longer time to completely fuse the grains of rice. You can make the Juk (Rice porridge) only with rice and water, and then add another grain, or nuts, vegetables, meat, fish and shellfish or medicinal herbs to the rice for a better taste.
3. Mieum, **Boembeok, Eungi**	**Mieum** A thin rice porridge made by pouring a large amount of water (over 10 times the amount of rice or grain) onto the grains and then bringing to the boil. You only drink the soup sifted from the mixture. **Beombeok** Main ingredients are corn, pumpkin and potatoes, and then soy beans, adzuki beans or grain flour are added when cooking it. **Eungi** Rice porridge made from the dried sediment of ground grains. Usually, Omija (from schizandra berries) juice is added to this porridge.
4. Guksu (Noodles) **Sujebi** (Dough Flakes Soup)	**Guksu** Is a food served in soup or mixed with other sauces. The dough of buckwheat flour, flour or potato powder is thinly sliced to make the noodle. **Sujebi** Small pieces of dough thinner than noodles are made and then put into a boiled clear soup and served. The clear soup is made from meat and anchovies etc.
5. Mandu (Dumplings)	Beef, chicken, tofu and bean sprouts are wrapped in a cover made of flour, buckwheat flour, vegetables or thinly cut fish fillet. The dumplings are then steamed, baked or put into a boiling meat broth or clear soup.
6. Tteokguk	To make white rice cakes, rice flour is first steamed and then pounded in a mortar. The white rice cakes are cut thinly and diagonally and put into boiling broth. Traditionally, roast pheasant was used to make the broth. These days, beef or chicken is used, while seafood such as oysters and anchovies may be used to make the broth. It is often garnished with stir-fried and chopped beef, fried egg strips and spring onions etc.
7. Others	Those foods that belong to the 'Main Dishes on the Meal Table' category, but not to any of the middle categories above.

Side Dishes

Middle categories	Definition
1. Guk (Soup)	**Jangguk (Clear soup)** A boiled chunky soup made by putting many solid ingredients into brisket beef soup and seasoned with soup soy sauce. **Tojangguk** A boiled chunky soup made by putting many solid ingredients into savory rice soaked in water and seasoned with soybean paste or red pepper paste. **Gomguk** A soup made by boiling many parts of meat and seasoned with salt. **Naengguk (Cold soup)** Boiled and then cooled-off water seasoned with soup soy sauce and then raw solid ingredients are added.
2. Jjigae, Jeongol (Stew, casserole)	**Jjigae** Stew made with 1:1 amounts of solid ingredients and soup. Jjigae is a thicker broth than guk (soup). Jjigae types are divided into Doenjang (soybean paste) jjigae, Gochujang (red pepper paste) jjigae and jeotguk (salted fish) jjigae etc. **Jeongol** An instantly boiled food made after pouring meat broth onto meat, fish, shellfish and vegetables.
3. Kimchi	**Kimchi** Fermented vegetables and seaweed preserved in salt, which is then mixed with jeotgal (salted fish) and added with seasonings including red pepper, spring onion, garlic and ginger. According to the main ingredients, Kimchi types are divided into those made from cabbage, radish, green vegetables, root vegetables, fruit and seaweeds, respectively.
4. Namul (Seasoned Vegetables and Salads)	**Saengchae** Raw vegetables preserved in salt and mixed with seasonings. **Sukchae** Parboiled vegetables mixed with seasonings or vegetables stir-fried in cooking oil with seasonings. **Others** There are others dishes that do not specifically belong to the Saengchae or Sukchae group, which are made of a variety of ingredients such as meat and vegetables.
5. Gui (Grilled Foods)	Grilled meat, fish and shellfish or deodeok (mountain herbs whose roots have a restorative effect) after being soaked in rich seasonings or salt.
6. Jorim, Jijimi (Braised and Fried Foods)	**Jorim** Braised meat, fish and shellfish or vegetables with strong seasonings, in that the ingredients have been sufficiently soaked with seasonings. Mainly seasoned with soybean sauce, yet mackerels or Pacific saury with a strong fish smell and red fillet should be braised using soybean paste and red pepper paste mixed with soybean sauce. **Jijimi** Jijimi is a thinner broth than Jjigae (stew) and thicker than Jorim (braised stew). Mostly, fish and shellfish are used as the main ingredients. Often, pan-fried jeons (pancakes, see below) are put into a small amount of broth.
7. Boggeum, Cho (Stir-fried Foods)	**Stir-fried** The generic name for foods that are stir-fried such as meat, fish and shellfish, vegetables, seaweed, grains and beans in cooking oil. These foods are either stir-fried only in cooking oil, or cooked using soybean sauce and sugar. **Cho** Foods such as Jeonbok (abalone) cho or Honghap (mussel) cho, which are highly braised with soybean sauce, sugar and cooking oil so that there would be no broth.

Side Dishes

Middle categories	Definition
8. Jeon, Sanjeok (Pan-fried Foods and Kebobs)	**Jeon** Chopped up or thinly sliced meat, fish and shellfish, vegetables or seaweeds are then seasoned with salt and pepper, finally both sides are pan-fried and covered with flour and egg. Also known as Jeonyueo, Jeonyuhwa or Jeonya. Thinly cut (about 1cm) ingredients (such as meat, fish and shellfish, vegetables or seaweed) of a length of 8~10cm, are then skewered on a stick differentiated by color. After those are covered with flour and egg, they are pan-fried in cooking oil.
9. Jjim, Seon (Steamed and Stuffed Foods)	**Jjim** Jjim indicates foods with large sliced ingredients with rich seasonings and then are boiled for a long time. The foods are steamed in vapor or warmed up in a double boiler. **Seon** Cucumber, pumpkin and tofu are slightly steamed together with other ingredients and soaked in soybean sauce mixed with vinegar.
10. Hoi (Raw and Slightly Cooked Foods, Rolls, Seafood Salads)	**Saenghoi** Thinly sliced or chopped up raw meat, fish, shellfish and seaweed served with red pepper paste mixed with vinegar, mustard sauce, salt or pepper. **Sukhoi** Slightly cooked fish and shellfish, vegetables or seaweeds and served. **Chohoi** Raw fish and shellfish, vegetables or seaweeds after being slightly seasoned with vinegar, soybean sauce or salt. **Ganghoi** The raw fish or meat ingredients are wrapped with thin vegetables such as parsley or small green onions, and then slightly soaked in red pepper paste mixed with vinegar. **Mulhoi** Thinly cut raw fish is mixed with seasonings, including spring onion, garlic and red pepper powder etc., by pouring water onto them this makes Mulhoi.
11. Mareun Banchan (Dry Dishes)	**Bugak** Fried laver. Thick and glutinous rice paste is applied to vegetables, fish and shellfish to dry them out to make fried laver. **Jaban** Jaban is available for serving after applying strong seasonings onto fish and shellfish and seaweed and preserving them for long time. **Tuigak** The main ingredient is seaweed without any other seasonings and this is cut and pan-fried in cooking oil. **Po** Seasoned meat, fish and shellfish after being spread out and dried.

Side Dishes

Middle categories	Definition
12. Sundae (Korean Sausage) **Pyeonyuk** (Pressed Meat)	**Sundae** Clotted pig blood, glutinous rice, parboiled bean sprouts and cabbage outer leaves are all mixed with rich seasonings, and then all of them are filled in a type of hog casing to be steamed just like a sausage. **Pyeonyuk** Brisket of beef or a shank of pork is boiled and cooled off under pushed pressure. Then thinly sliced and served.
13. Muk, Dubu (Jelly and Tofu)	**Muk** This is starch jelly made from buckwheat, mung beans, acorns or arrowroot starch powder plus water which are then all boiled and cooled off into a jelly. **Dubu** To make tofu, beans are soaked in water and then ground. After biji (pureed soybean) is filtered by being boiled, it turns into a solid tofu with the addition of bitterns (salty water) as a coagulant.
14. Ssam (Wrapped foods)	Rice and side dishes are wrapped in vegetables or seaweeds. The ingredients can be either raw or cooked.
15. Jangajji (Pickles)	These foods are made by putting vegetables into salty water, soybean sauce, soybean paste or red pepper paste and then being allowed to ferment.
16. Jeotgal, Sikhae (Salted and Fermented seafood)	**Jeotgal** Fermented foods made by self-degradable enzymes and beneficial microorganisms and putting salt into fillets, intestines and eggs of fish and shellfish. Salt used should be about 20% of the total weight of the fish. **Sikhae** Fermented foods made from salt-preserved fish fillets mixed with rice (millet rice and rice), radish strips, red pepper powder and other seasonings.
17. Jang (Sauce)	Basic Jang types are soybean sauce, soybean paste and red pepper paste, all of which are made from the blocks of fermented soybeans.
18. Others	Those foods that belong to the 'Side Dishes' category, but not to any of the middle categories above.

Rice Cakes

Middle categories	Definition
1. Jjintteok (Steamed)	Also known as 'Sirutteok', which is tteok (rice cake) made from grain flour and a garnish and steamed together in steamer.
2. Chintteok (Pounded)	A rice cake which is made by pounding rice in a mortar after cooking the rice or steaming the grain flour.
3. Jijintteok (Pan-fried)	Grain flour is made into dough to make shapes, and then pan-friend in cooking oil on a frying pan.
4. Boiled tteok	Grain flour is made into dough to make shapes, and then steamed and coated with crumbs.
5. Others	Those foods that belong to the 'Rice Cakes' category, but not to any of the middle categories above.

Snacks

Middle categories	Definition
1. Yumilgwa	These foods are made by frying the shaped dough mixed with flour, honey and cooking oil, then coating them with honey or starch syrup.
2. Yugwa	These foods are made by steaming the dough of glutinous rice flour mixed with bean juice or liquor, and kneading the dough and thinly pressing it to make it dry and finally frying it in cooking oil to be coated with crumbs and sugar.
3. Dasik	A generic name for the foods made from honey dough mixed with grain flour, medicinal herb powder, nuts and edible pollens and then made in a honey cake mold.
4. Jeonggwa	Braised and candied foods in honey or sugar syrup, using the chunks or slices of plant roots, stems or fruits.
5. Yeotgangjeong	Sliced sweets made of beans, sesame seeds, or nuts mixed with malt liquid, grain syrup, honey or sugar syrup.
6. Dang (hard taffy)	The generic name for foods that cover braised rice, glutinous rice and sorghum or sweet potato in a sweet malt.
7. Others	Those foods that belong to the 'Gwajeong' category, but not to any of the middle categories above.

Beverages

Middle categories	Definition
1. Tea	After pounding a variety of medicinal herbs, fruits or tea leaves into powder, drying them or thinly cutting and soaking them in honey or sugar syrup, tea can be made by putting these into boiling water.
2. Hwachae	Korea's fruit punch using fruits and flowers sliced in various forms and marinated in honey or sugar. Otherwise, these are put into Omija (schizandra berries) juice, sugar water or honey water
3. Sikhye	Korea's traditional beverage made with cooked glutinous rice or non-glutinous rice which is then put into powdered malt brewed water and then fermented for a specific time and at a certain temperature.
4. Sujeonggwa	Korea's traditional beverage made of water boiled down with ginger and cinnamon with the addition of dried persimmons and added honey or sugar for a sweet taste.
5. Others	Those foods that belong to the 'Beverages' category, but not to any of the middle categories above.

Liquor

Middle categories	Definition
1. Yakju & Takju	Beverage containing alcohol with fermented grains.
2. Hard liquor	Liquor such as Soju (Korean distilled spirits) made by distilling grain-fermented liquor again, so that it contains a much greater amount of alcohol.
3. Others	Those foods that belong to the 'Rice Cakes' category, but not to any of the middle categories above.

2

[Seoul · Gyeonggi]

Seoul became the capital city starting from the era of the Joseon Dynasty and the culture of the royal family significantly influenced the regional foods of this area. This was a place where the local foods were brought in by people who had arrived from each region in order to pay homage to the royals and moreover was a place where all the various regional foods of the kingdom were consumed in the same place. Therefore, food from Seoul, in terms of the ingredients used, is typically more diverse than food from any other region.

Food from Seoul is neither salty nor spicy. Instead, food here is simply seasoned appropriately. Although the amount of food in the dishes could be regarded as rather small, there is always a wide variety dishes made available.

In the case of kimchi, refreshing salted fish, such as salted shrimp, salted yellow corvina, salted croaker, raw shrimps and raw hairtail, were used abundantly, offering a fresh and delicious taste. Food from Seoul was decorated stylishly with five-color garnishes and offered an elegant taste since many foreign envoys visited the area.

Compared to Seoul, the local food from the wider Gyeonggi region is somewhat humble, and the seasoning is simple as well, but it still offers a significant amount of choice. Just like pumpkin porridge or sujebi (dough flakes soup), there is a considerable amount of food made from pumpkins, potatoes, maize, flour, red bean and so forth.

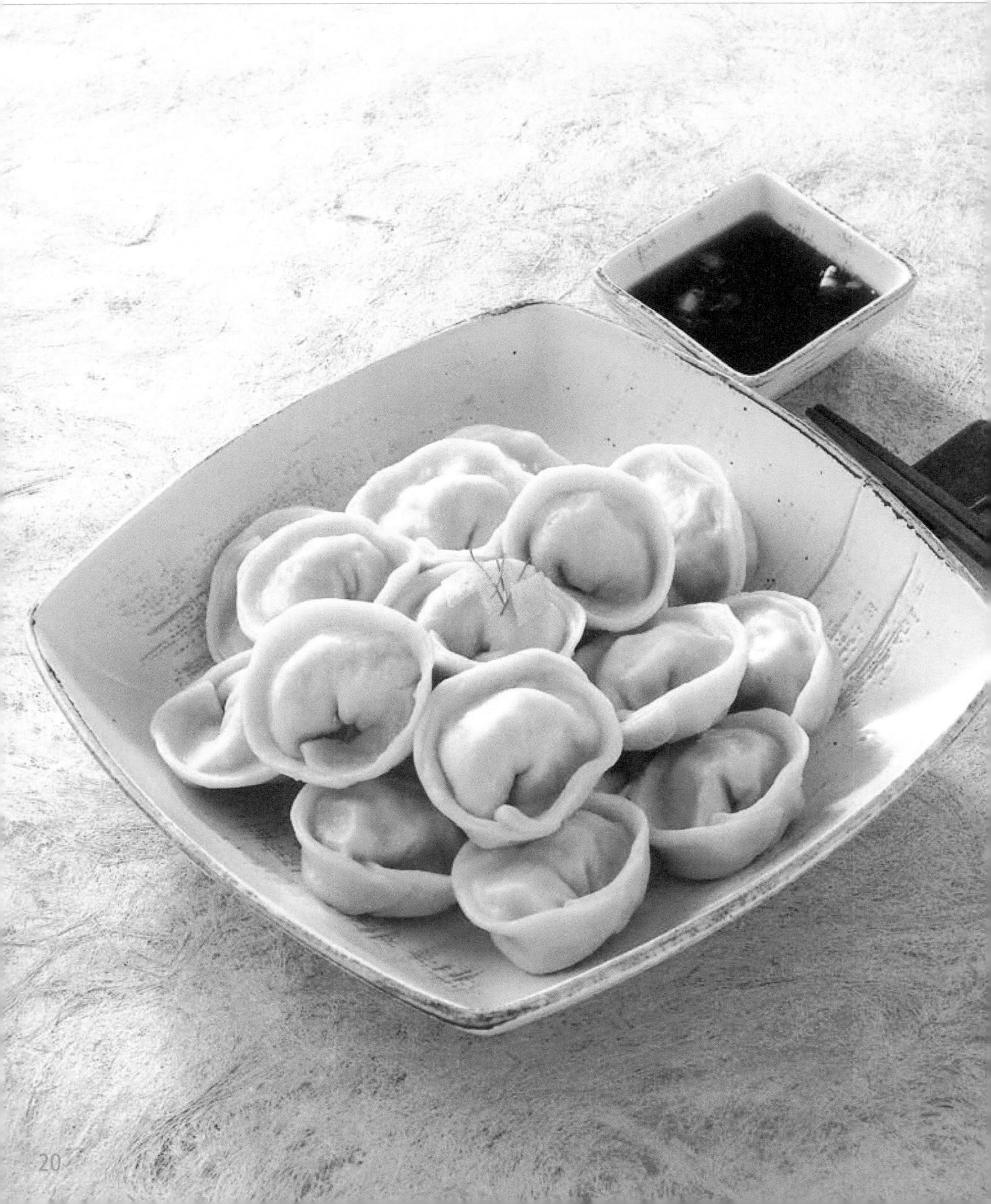

Gaeseong-style pyeonsu
(Dumplings)*

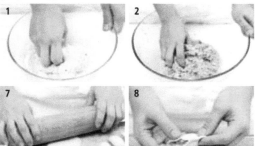

Ingredients

For the dumpling skins/wrappers flour 220g, one egg white, water, salt 1 teaspoon
For the dumpling filling beef 100g, pork 100g, tofu 150g,
mung-bean sprouts 100g, kimchi 100g, one egg yolk, a pinch of salt
For the seasoning for the meat soy sauce 1 tablespoon, chopped green onions 2 tablespoons, crushed
garlic 1 tablespoon, sesame oil 2 tablespoons, sesame salt 1 tablespoon, black pepper 1 teaspoon
For the seasoning for the dumpling filling salted shrimps 1 tablespoon,
red hot pepper powder 1 teaspoon, sesame oil 1 tablespoon, a pinch of salt

Cooking Directions

1 For the dumpling skins/wrappers, in which the fillings will be placed, combine the flour
  and salt in a bowl and sieve them. Add the egg white and water to the flour and beat
  until the mixture turns into a dough form. Wrap the dough in a wet cotton cloth for 30 minutes.
2 Finely mince the beef and pork. Season the meats with the meats' seasoning ingredients.
3 Wrap the tofu in a cotton cloth and press with a heavy object so as to finely crush it.
  Combine the tofu, the chopped salted shrimps and filling seasonings.
  Mix them thoroughly until the color turns pink.
4 Parboil the mung-bean sprouts in salty boiling water.
  Squeeze out the excess water and finely chop up the mung-bean sprouts.
5 Squeeze out the excess juice of the kimchi and chop it into small pieces (0.5 cm)
6 For the filling of the dumplings, mix the meat, tofu, mung-bean sprouts
  and kimchi with the egg yolk and salt.
7 Roll out the dough (from point 1) into circles of dough with 0.3 cm thickness and 6 cm diameter.
8 Spoon out a lump of the dumpling filling and place it into the middle of the dough.
  Fold the dough with the filling in half. Pinch the top of the semi-circle parcel together
  and seal all the edges of the dumpling. Drop the dumplings into a big pot of boiling water
  until they float to the surface. Rinse them in cold water, drain and put aside in a bowl.
9 Place the boiled dumplings on a plate. Serve with vinegar or soy sauce.
  Alternatively, put the dumplings into some boiling beef soup for some time
  and garnish with the chopped beef and fried egg strips.

Note

'Mandu' came to Korea from China, and spread into the northern region of Korea. Even today, those in the southern regions do not cook them often. In the Goryeosa (History of Goryeo), there is mention of someone who was punished after breaking into a kitchen to steal mandu during the 4th year of King Chunghye's reign. Given this, it is possible to know that there were mandu being eaten in Korea at least during the Goryeo Era as well. In Gyeonggi and Seoul, mandu was referred to as pyeonsu since mandu is boiled in water before being eaten.

---

* Dumplings are also familiar to people in English-speaking countries. The Korean-style dumplings as well as dim sum (China) or kyoja (Japan) are also well received overseas.

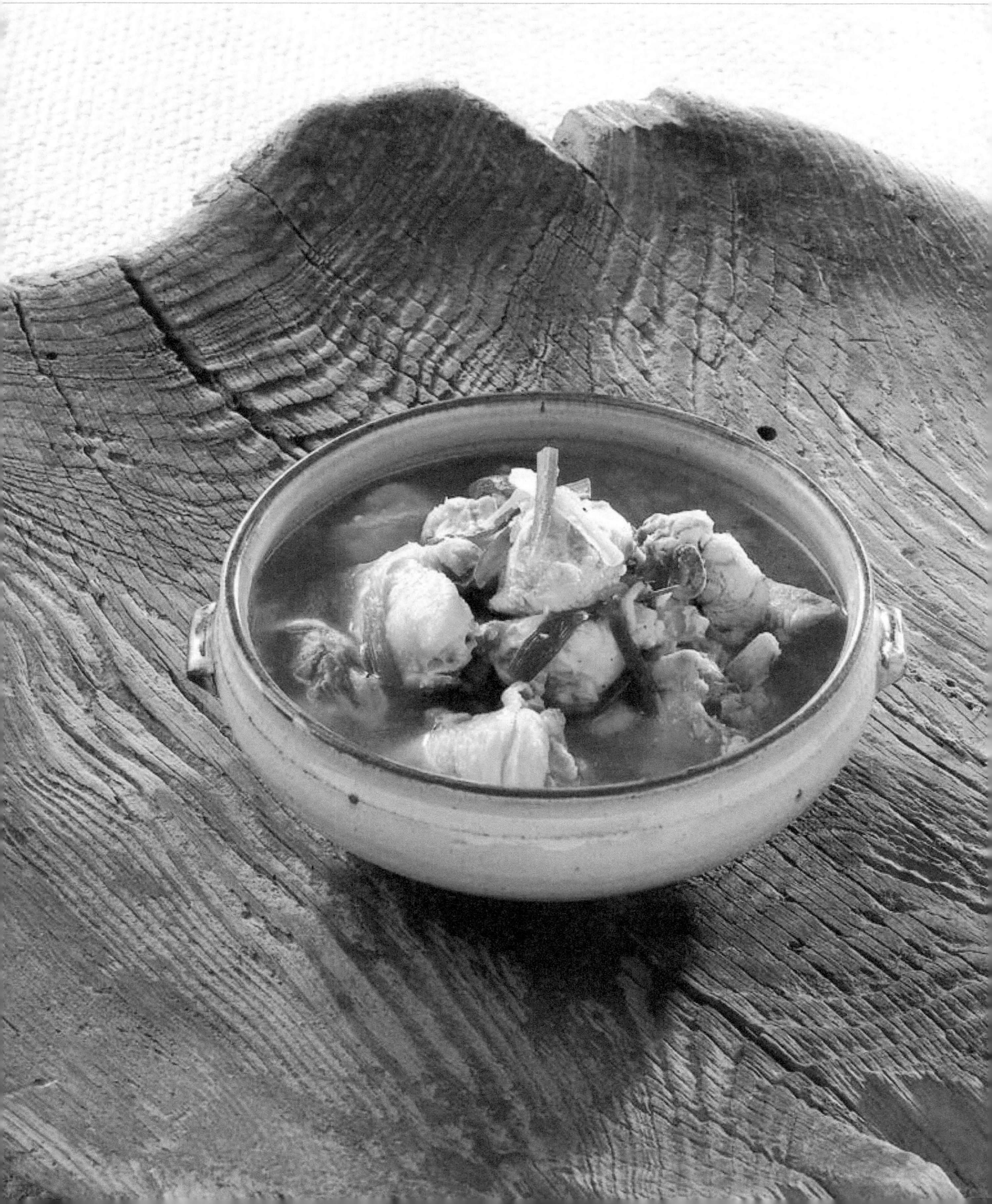

Dakjeot guk
(Chicken Soup with Salted Shrimps)

Ingredients

Chicken 500g, some sesame oil

For the seasoning sauce chopped green onions 3 tablespoons, crushed garlic 1 tablespoon, ginger juice 1/2 tablespoon, salted shrimps 1 tablespoon, sesame oil 3 tablespoons, black pepper

For the broth water 5L, salted shrimps 2 tablespoons

Cooking Directions

1 Cut the chicken below the belly and remove the innards and grease. Cut it into some pieces (4-5cm). Parboil the chicken pieces in boiling water and rinse well in cold water.

2 Combine the seasoning ingredients to make the seasoning sauce.

3 Marinate the boiled meat in the seasoning sauce for an hour.

4 Heat up a pan and line it with sesame oil. Stir-fry the marinated chicken. When the color of the meat surface turns white, add the water and salted shrimps and cook over a strong heat.

5 When it starts to boil, reduce the heat and simmer for another 30 minutes. Add the salted shrimps.

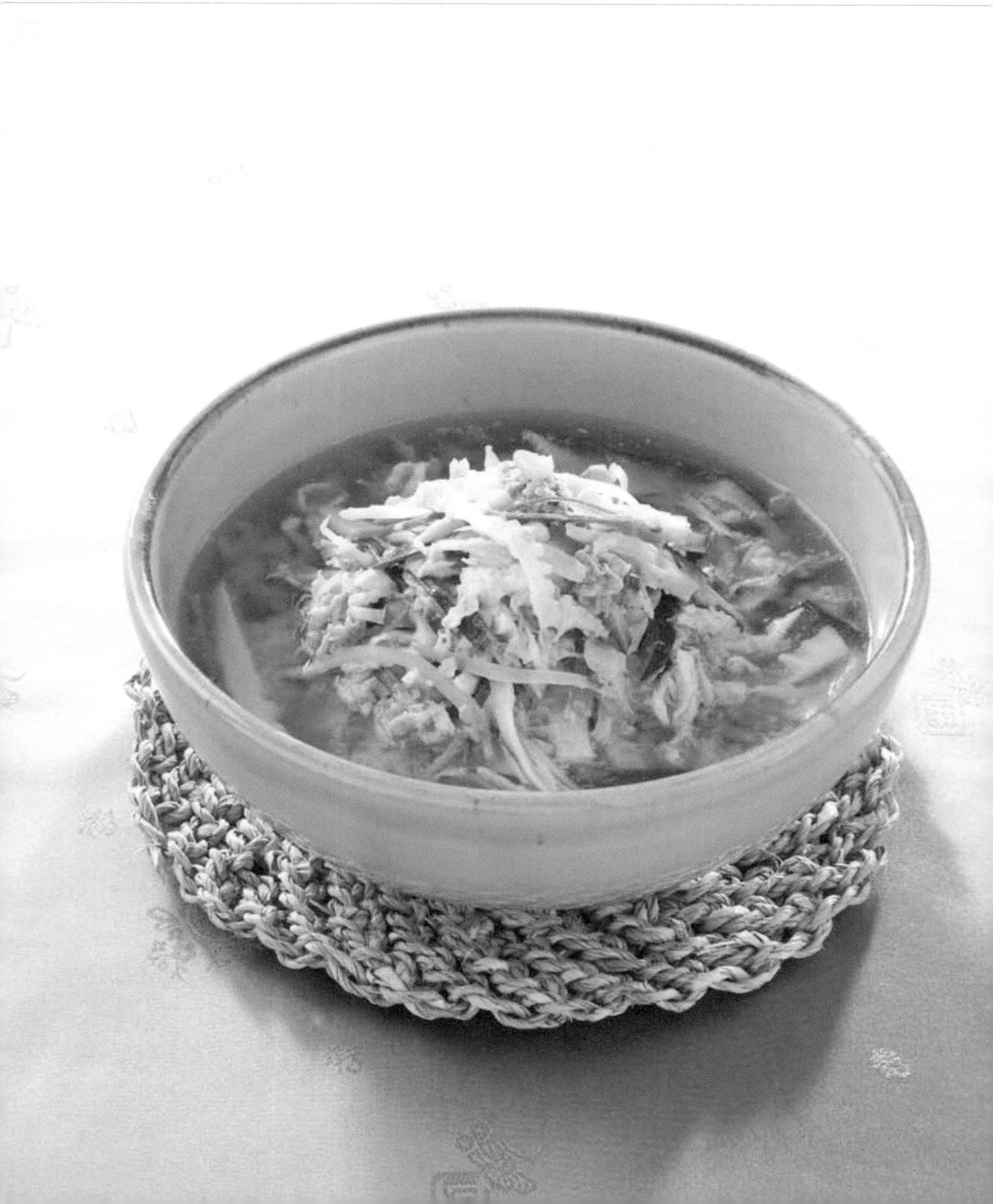

Chogyo tang
(Summer Chicken Soup)

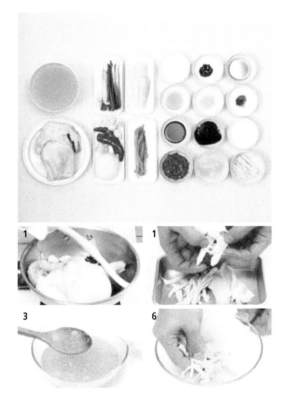

Ingredients

Chicken 1kg, bellflower roots 80g, water parsleys 50g, bamboo shoots 100g, flour 110g, eggs 100g, small green onions 30g, red hot peppers 10g, beef 100g, dried shiitake mushrooms 10g, chicken broth 2L, sesame oil 1/2 tablespoon, light soy sauce, anchovy sauce, salt, a pinch of ground black pepper

For the chicken broth water 2.6L, ginger 20g, garlic 30g, onions 100g

For the seasoning for the beef and shiitake mushrooms soy sauce 1 tablespoon, chopped green onions 1 tablespoon, crushed garlic 1/2 tablespoon, sugar 1/2 tablespoon, sesame oil 1/2 tablespoon, a pinch of ground black pepper

For the seasoning for the chicken flesh salt 1 teaspoon, chopped small green onions 1 tablespoon, crushed garlic 1/2 tablespoon, sesame oil 1/2 tablespoon, ginger juice 1 teaspoon, a pinch of ground white pepper

Cooking Directions

1 Clean the chicken. Boil it with the ginger, garlic and onions in water and take out them.
Use the water as the chicken broth. Peel the skin and tear off the flesh from the bone.
Put the flesh aside. Put the bones into the chicken broth.
Boil the chicken broth again, skimming off the grease. Take out the bones to keep the broth clear.

2 Shred the bellflowers thinly and knead with salt to eliminate the bitter taste and soften the texture.
Clean and cut the water parsleys into 3cm pieces and parboil in water. Slice the bamboo shoots thinly, parboil, and then fry. Cut the red hot peppers into strips of 3x0.3x0.3cm

3 Mix and season the chicken flesh, bellflower roots and water parsleys with the chicken flesh seasoning.

4 Chop the meat. Soften the shiitake mushrooms in water and slice into 0.3cm thick strips.
Mix with the beef and shiitake mushroom seasoning.

5 Add the flour and eggs to all the seasoned ingredients and knead well.
Add small green onions and mix well to make dough.

6 Season the broth lightly with soy sauce, anchovy sauce, and salt. Then add a spoonful of the dough one by one into the seasoned broth. Turn off the heat when the dough floats to the surface.
Sprinkle with sesame oil and ground black pepper.

Dwaeji galbi kong biji jjigae
(Pork Ribs and Pureed Soybean Stew)*

Ingredients

Soybeans 160g, anchovy broth or dried Pollack broth 800 ml, pork ribs 600g,
soft cabbage (boiled cabbage) 500g, sliced radishes 150g
For the seasoning: ginger juice 1 teaspoon, anchovy sauce 1 tablespoon,
crushed garlic 1 tablespoon, chopped green onions 2 tablespoons, sesame oil 1 tablespoon,
some ground black pepper, maesil (Japanese apricot) syrup 1 tablespoon
For the seasoning for the soft cabbage and sliced radishes: crushed garlic 1 tablespoon,
chopped green onions 2 tablespoons, sesame oil 1 tablespoon, salted shrimps 1 tablespoon,
some ground black pepper
For the seasoning soy sauce: soy sauce 3 tablespoons, thick red hot pepper powder 1 tablespoon,
chopped small green onions 2 tablespoons, crushed garlic 1 tablespoon, sesame oil 1 tablespoon,
sesame salt 1 tablespoon, maesil (Japanese apricot) syrup 1 teaspoon
For the salted shrimp sauce salted shrimps 2 tablespoons, chopped small green onions 2 tablespoons,
thick red hot pepper powder 1 tablespoon, crushed garlic 1 tablespoon, sesame oil 1 tablespoon,
sesame salt 1 tablespoon, maesil (Japanese apricot) syrup 1 tablespoon

Cooking Directions

1 Soften the soybeans overnight. Pour water over the softened soybeans and then finely grind.
2 Parboil and chop the soft cabbage into small pieces. Combine the cabbage and the sliced radishes
  and then season.
3 Parboil the pork ribs and marinate them in the seasoning sauce for an hour.
  Stir-fry the marinated ribs and soft cabbage.
4 Pour the anchovy broth over the stir-fried ribs and soft cabbage from point 3 and boil thoroughly.
5 When the pork ribs are cooked, pour the ground soybean porridge over the pork ribs
  and cook over a low heat.
6 Serve with the salted shrimp sauce or seasoning soy sauce.

---

* In general, not-too-spicy meats are preferred when preparing this dish.

Byeongeo gamjeong
(Thick Butterfish Stew)*

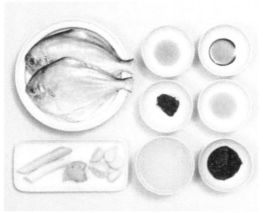

Ingredients

Butterfish 480g

For the broth Anchovy broth 200ml, red hot pepper paste 3 tablespoons, soybean paste 1/2 tablespoon, anchovy sauce 1/2 tablespoon, light soy sauce 1/2 tablespoon

For the seasoning sliced green onions 2 tablespoons, sliced garlic 1 tablespoon, sliced ginger 2 tablespoons, sesame oil 2 teaspoons

Cooking Directions

1 Remove the scales and innards from the butterfish and score the flesh.
2 Finely slice the green onions, garlic, and ginger and mix with the other seasonings.
3 Put the red hot pepper paste, soybean paste, anchovy sauce, and light soy sauce into the anchovy broth and bring them to the boil. Place the butterfish into the broth and braise.
4 Add the seasonings from point 2 above to the butterfish and keep boiling the mixture down.

Note

Gamjeong (thick stew) has less broth than jjigae (stew), which is ideal for the topping sauce for ssam (vegetable leaf wrap). Butterfish can be replaced with yellow corvina.

* A thick fish stew is more welcomed than a normal stew dish.

Kkwong kimchi
(Pheasant Kimchi)*

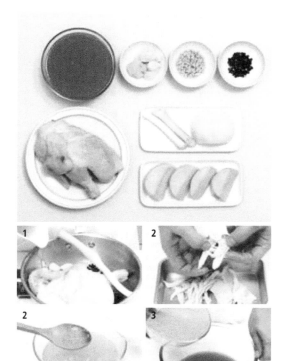

Ingredients

One pheasant, radishes for dongchimi (cold radish kimchi broth) 1kg, water 2L, dongchimi broth 2L, pine nuts 1 tablespoon, green onions 35g, onions 80g, ginger 10g, garlic 30g, some black pepper

Cooking Directions

1 Remove the innards of the pheasant and rinse thoroughly. Add the pheasant with the green onions, onions, ginger, garlic, and black pepper into a pot and boil thoroughly.

2 Tear off the meat and shred the flesh. Cool the broth and skim off the grease.

3 Mix the dongchimi broth with the pheasant broth.
Add the sliced radishes, pheasant meat and pine nuts.

Note

Kkwong kimchi was mentioned in 'The Umsik Dimibang' written by a lady called Mrs. Jang from Andong in 1670. According to the book, it was also called 'Saengchi Chimchae,' which was made by putting the ingredients into warm water with salt. It was made and

* You can use only the flesh of the meat.

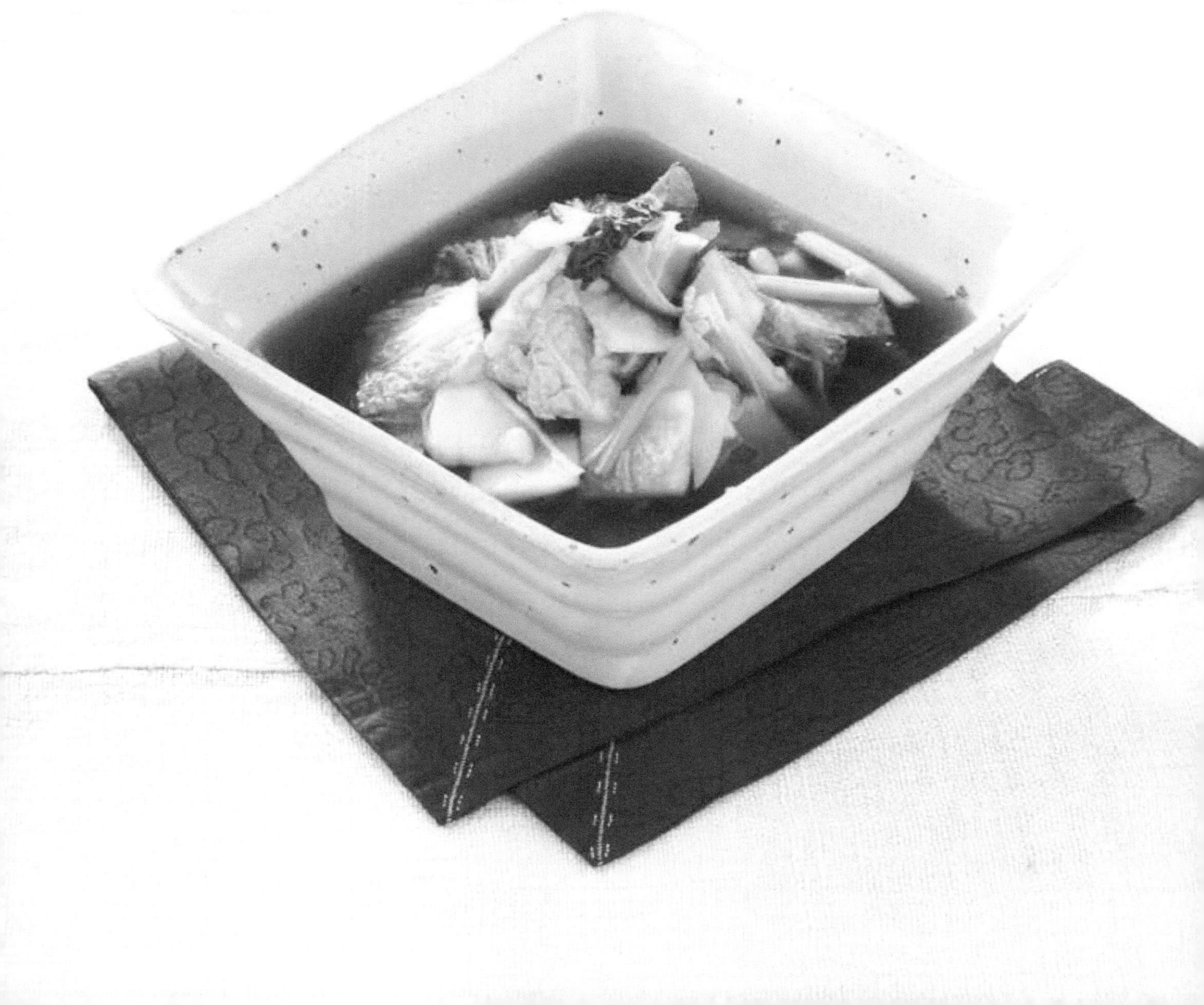

Jang kimchi
(Kimchi in Soy Sauce)*

Ingredients

Chinese cabbage 400g, radishes 150g, water parsleys 100g, mustard leaves 150g,
small green onions 50g, shiitake mushrooms 10g, manna lichen mushrooms 3g, chestnuts 100g,
jujubes 20g, persimmons 140g, pears 370g, garlic 30g, ginger 10g, pine nuts 1 tablespoon,
mature soy sauce 1/2 cup
For the broth soy sauce 1/2 cup, water 1.2L, honey (or sugar) 3 tablespoons

Cooking Directions

1 Remove the outer leaves of the Chinese cabbage.
   Cut off every leaf and cut them into pieces (3.5 cm x 3 cm)
2 Choose radishes that feel firm and avoid those that look wilted or are soft.
   Clean them thoroughly and cut into pieces smaller than the Chinese cabbage.
3 Pour the mature soy sauce over the Chinese cabbage and radishes and stir well.
   Soak them in the sauce for a while.
4 Nip off the tails of the mustard leaves and water parsleys and clean them.
   Cut the stems into 3.5cm long pieces. Soften the shiitake mushrooms in water, then trim and shred.
   Trim and wash the manna lichen mushrooms, and cut into pieces 0.2cm thick.
5 Slice the chestnuts into 0.3 cm thick pieces.
   Deseed the jujubes and cut lengthwise into about 3 pieces.
6 Peel the persimmons and pears, and cut into pieces of the same size as the radish.
7 Choose only the white part of the small green onions, cut it into 3.5cm long pieces.
   Finely slice the garlic and ginger
8 Nip off the peaks of the pine nuts and wipe with a dry cloth.
9 Combine all the prepared ingredients and mix well, pour the mature soy sauce made from the
   radishes and Chinese cabbage (see point 3) and keep it as it is for a day.
   Then, pour the prepared broth over it. Keep it until it is fermented.

Note

Jang kimchi features the sweet and savory radishes and Chinese cabbage that are seasoned with soy sauce. As it matures fast, it is difficult to make large quantities at a time. When the weather is cool, it becomes best to eat in 4 to 6 days and when the weather is hot, it is better to eat in 2 days. Jang kimchi tastes better in autumn and winter. Several pine nuts are topped on the food when it is served. Jang kimchi is also served for a large dining table. It also goes well with tteok guk (rice cake soup).

---

* As it is not so spicy and contains fruits it is enjoyed by many people.

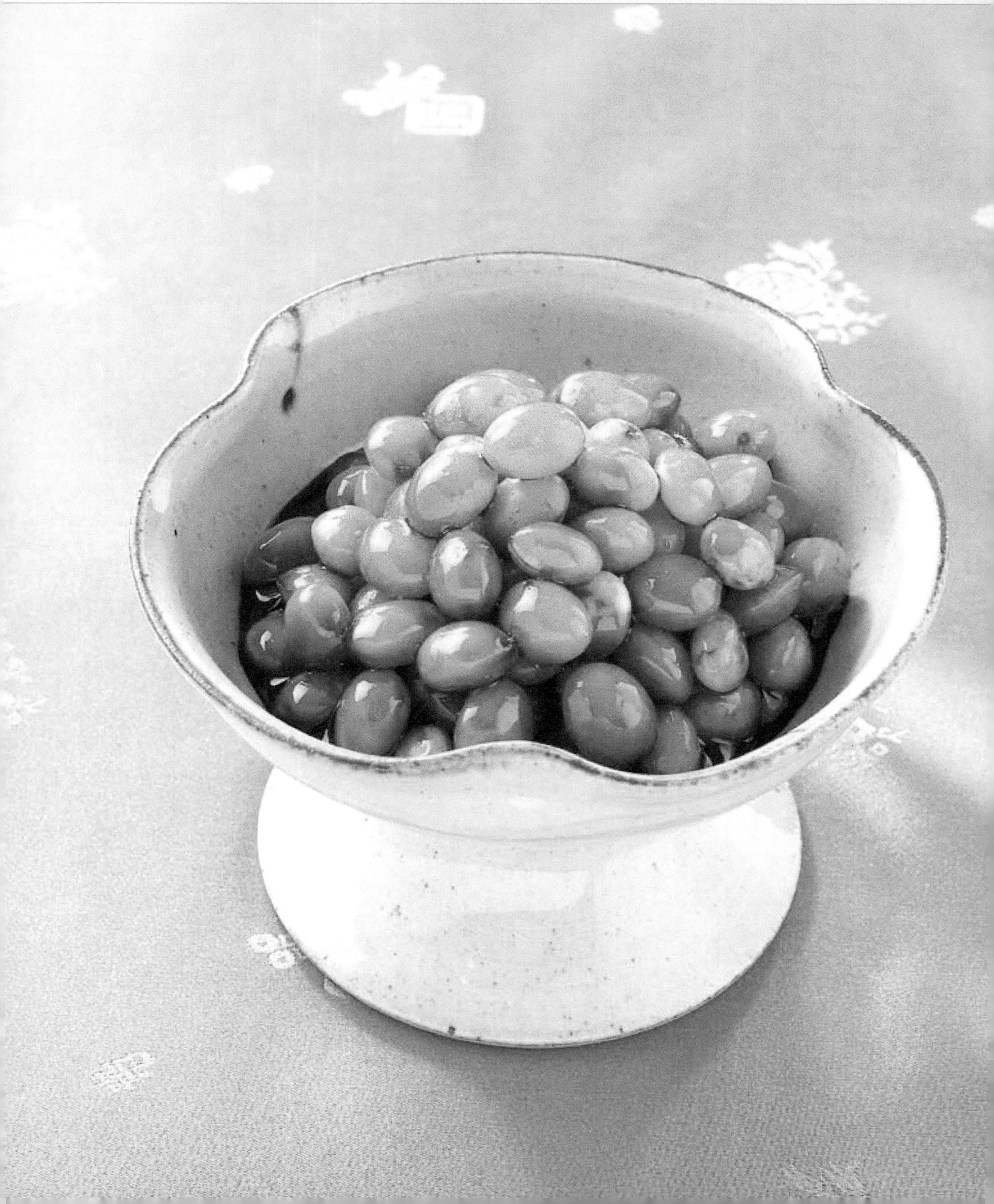

Eunhaeng jangjorim
(Gingko Nuts Braised in Soy Sauce)

Ingredients

Gingko nuts 500g, cooking oil 1/2 tablespoon
For the seasoning soy sauce 1/2 cup, dextrose syrup 1/3 cup, sugar 1/3 cup,
refined rice wine 3 tablespoons, water 100ml, some sesame oil

Cooking Directions

1 Rinse the gingko nuts with water, stir-fry and peel the skin.
   Wipe the remaining grease with a clean cloth.
2 Put the soy sauce, dextrose syrup, sugar, water, and refined rice wine in a bowl and bring to the boil.
   When the sauce is reduced by half, add the prepared gingko nuts.
3 Reduce the heat. When the gingko nuts are cooked and look glossy, turn off the heat.
4 Sprinkle over some sesame oil before eating.

Dubujeok
(Pan-fried Tofu with Pork)*

Ingredients

Tofu 1kg, pork 150g, some cooking oil
For the tofu seasoning salt 1 teaspoon, some black pepper, starch 2 tablespoons
For the pork seasoning soy sauce 1 tablespoon, sugar 1/2 tablespoon,
chopped green onions 1 tablespoon, crushed garlic 1/2 tablespoon,
ginger juice 1 teaspoon, some black pepper
For the vinegar soy sauce soy sauce 1 tablespoon, vinegar 1/2 tablespoon,
maesil (Japanese apricot) syrup 1 tablespoon, mineral water 1 tablespoon

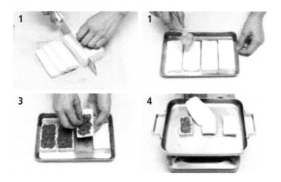

Cooking Directions

1 Press the tofu slightly to wring out the water, cut into 7mm thick pieces.
   Season it with salt and black pepper. Coat it with starch.
2 Season the minced pork with the prepared ingredients.
3 Spread the minced pork batter onto the one side of each tofu thinly and evenly.
4 Line a pan with cooking oil. Fry one side of the minced meat and then the other side.
5 Serve with the vinegar soy sauce.

Note

Tofu is known to have been invented by Prince Liu An of the Han Dynasty, around the 2nd Century B.C. It is presumed that it was brought into Korea during the Tang Dynasty. In ancient times, tofu was called po (泡). During Korea's Joseon Dynasty period, the temple that was responsible for making tofu was called Joposa (造泡寺).

* In the US or the UK, tofu is catching on in popularity. However, they prefer the hard tofu to the soft one.

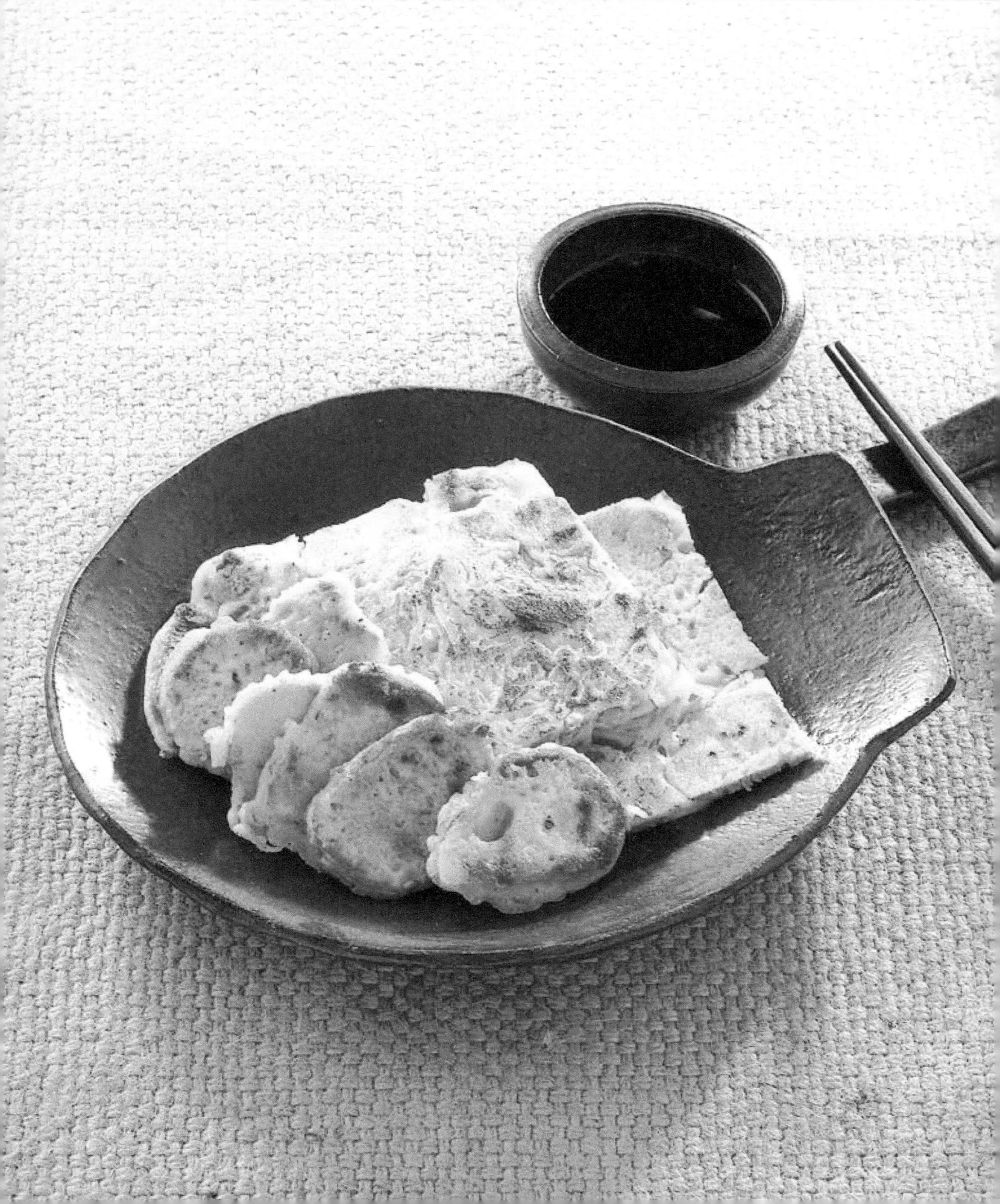

Jeyuk jeonya
(Pan-fried Pork with Flour)

Ingredients

Pork (shank)600g, flour 110g, cooking oil, water, salt 1/3 teaspoon

Cooking Directions

1. Boil the pork and place in a bowl. Gently squash down the boiled pork slightly and then slice it.
2. Combine the flour with the water and salt and mix well.
3. Put some oil into a heated pan and then scoop the batter from point 2 and drop it into the pan. Place the sliced meat pieces onto the batter and lightly cover with another scoop of batter. Pan-fry both sides until it turns a beautiful yellow.
4. Cut up the pan fried pork into bite sizes. Place the pieces of pork in a dish ready for being placed on the meal table.

Tteok jjim
(Steamed Rice Cake)

Ingredients

Garae tteok (sticks of round rice cake) 500g, beef shank 200g, brisket 200g, shredded beef 100g, radishes 100g, carrots 100g, dried shiitake mushrooms 15g, water parsleys 50g, eggs 50g, gingko nuts 20g, broth (made from brisket) 200ml

For the seasoning for the beef shank and brisket

soy sauce 1 tablespoon, chopped green onions 1 tablespoon, sugar 1/2 tablespoon, crushed garlic 1/2 tablespoon, black pepper, sesame oil 1 tablespoon

For the seasoning for the shredded beef

soy sauce 1 tablespoon, chopped green onions 1 tablespoon, crushed garlic 1/2 tablespoon, sugar 1/2 tablespoon, black pepper, sesame oil 1 tablespoon

For the seasoning sauce to be served with the dish soy sauce, sesame salt, sugar, sesame oil

Cooking Directions

1 Boil the brisket and beef shank thoroughly and place in a bowl.
   Cut the meat into big pieces and season them.
2 Shred the softened shiitake mushrooms and season along with the shredded beef.
3 Cut the garae tteok into 5cm long pieces and then cut each piece again into 4 ones.
   If they become hardened, parboil them in boiling water.
4 Parboil the radishes and carrots and cut into pieces of the same size as the garae tteok.
   Cut the water parsleys into 4cm pieces. Peel both the outer and inner shell of the gingko nuts.
5 Fry the eggs individually in a pan into very thin layers. Cut them into 2cm wide diamond shapes.
6 Stir-fry the seasoned beef strips and shiitake mushrooms. Add the seasoned beef shank, brisket, radishes and carrots, and then pour the broth over the ingredients and simmer.
7 When the amount of the broth is reduced by half, add the garae tteok and gingko nuts, mix well and season.
8 Add the water parsleys just before removing the pan from the heat.
   Place the cooked food in a bowl, and then top with the white and yellow egg strips.

Susam ganghoi
(Rolled Fresh Ginseng)*

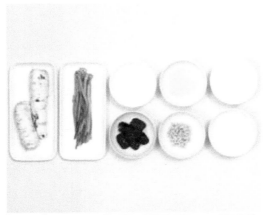

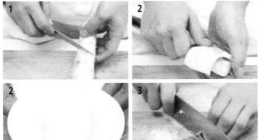

Ingredients

Fresh ginseng 5 roots, jujubes 10g, water parsleys 5 strips, sugar 1 tablespoon, vinegar 1 tablespoon, salt 1/2 teaspoon, honey 2 tablespoons, some pine nuts

Cooking Directions

1 Choose some mid-sized roots of fresh ginseng. Wash them thoroughly.
2 Cut the two roots of the fresh ginseng into 4cm long pieces.
  Marinate them in water mixed with the salt, sugar and vinegar
3 Deseed the jujubes and slice them.
4 Put the sliced jujubes onto the fresh ginseng pieces and roll them up.
  Top with the other little jujubes for decoration.
5 Cut the other two fresh ginseng roots into 1cm x 3.5 cm pieces.
  Top with the sliced jujubes. Tie them together with the parboiled water parsley strips.
6 Serve with honey, vinegar or a hot pepper paste.

---

* Fresh ginseng is well known as being a healthy food. Though it tastes bitter, it remains well received by many people.

Yongin oiji
(Yongin-style Cucumber Pickle)*

Ingredients

Yellowish overripe cucumbers (Nogak) 2kg, clear rice water 18L, salt 4kg, some water

Cooking Directions

1. Wash and place the cucumbers in a jar. Press them down with some pebbles.
2. Pour cold water into the jar. Add the salt to the water and stir. Leave overnight.
3. Empty out the water from the jar into a pot. Boil the water and then leave it to cool. Then pour the water into the jar again. Repeat this process three or four times. Then the pickled cumbers are ready for serving, chopped or whole.

Note

In the Buin Pilji, a cook book which can be traced back to the Joseon Dynasty period, includes records on the preservation of cucumbers produced in Yongin.

Moreover, this is also included in the Gyuhap Chongseo which indicates that this is a traditional food with history. This was a type of kimchi with a crispy sensation and sweet and sour taste that may have developed into the oiji (cucumber pickle) of today.

---

* This pickle is similar to those in the West. The European pickles however, are salty and don't taste so sweet.

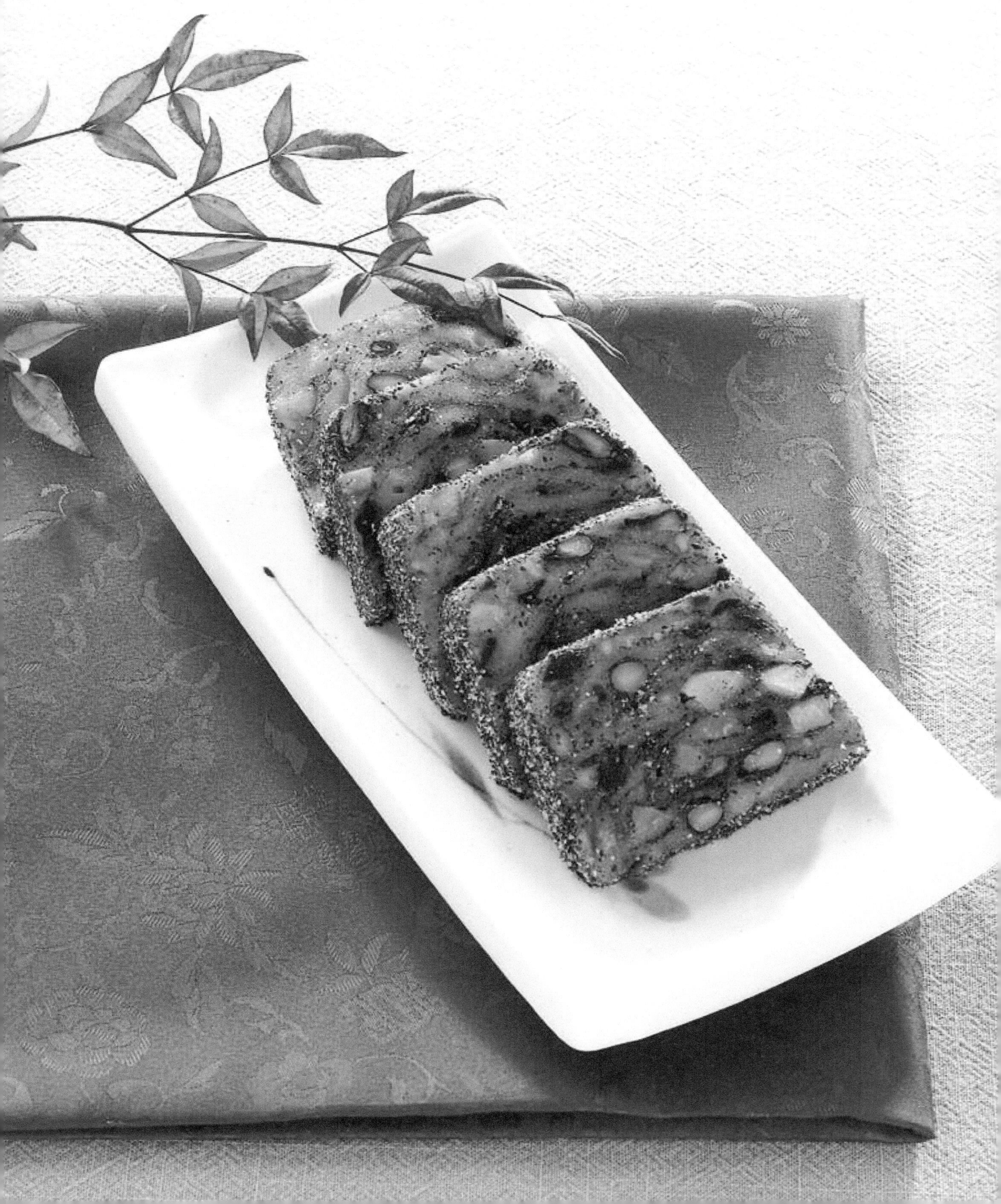

Gureum pyeon
(Gureum Rice Cake)

Ingredients

Glutinous rice powder 1kg, red bean powder 1 cup, sugared water 100ml, jujubes 100g, chestnuts 200g, walnuts 40g, kidney beans 1/2 cup, pine nuts 35g, water, some sugar

Cooking Directions

1. Mix the glutinous rice powder with water.
2. Add the water to the red bean powder and bring to the boil. Then pour it through a tray to drain the water. Stir-fry the boiled red bean to eliminate the remaining liquid.
3. Slightly steam the chestnuts and peel off their skins. Deseed the jujubes and cut them into 2 or 3 pieces. Boil the softened kidney beans.
4. Wipe the pine nuts with a dry cotton cloth. Peel the inner shell of the walnuts. Cut the walnuts into 2 pieces.
5. Combine all the ingredients from point 3 and 4, and braise them in the sugared water.
6. Mix well the ingredients from point 5 with the glutinous batter from point 1. Steam them in a pot to make glutinous rice cakes.
7. Sprinkle red bean powder in a square mold. Take some pieces from the glutinous rice cakes, cover it with the red bean powder and press it into the mold by hand. Place a heavy pot on the mold for 2 or 3 hours to make into a beautiful shape.
8. Takeout the rice cake from the mold and cut into bite sizes.

Duteop tteok
(Rice Cake with Red-beans and Nuts)

Ingredients

For the tteok glutinous rice 500g,
soy sauce 1 1/2 tablespoons, sugar 3 tablespoons, honey 3 tablespoons
To make the powdered red bean mixture de-husked red beans 4 cups,
mature soy sauce 2 tablespoons, sugar 4 tablespoons, honey 5 tablespoons,
cinnamon powder 1/2 teaspoon, black pepper
For the filling of the tteok chestnuts 100g, jujubes 50g, walnuts 40g, pine nuts 25g,
citron pieces in citron syrup 1/2 tablespoon, citron syrup 1 tablespoon

Cooking Directions

1 Clean the glutinous rice well and soften in water for six hours.
Then drain the water and leave the rice to stand for 30 minutes.
Grind the dried glutinous rice until it becomes like flour.

2 Add the soy sauce into the glutinous rice powder and mix well.
Sieve them and then add the sugar and honey to make the tteok.

3 Soften the red beans in water. De-husk and wash the red beans well.
Lay a wet cotton cloth inside a steaming pot and then place in the red beans.
Steam them thoroughly.

4 Pour the steamed red beans into a big bowl. Grind them slightly and sieve them.

5 Add the soy sauce, sugar, honey, cinnamon powder and black pepper to the sieved red beans
and mix well. Pan-fry them slightly and sieve them again.

6 Chop off the tails of the pine nuts. Cut the chestnuts and jujubes into pieces the same size
as the pine nuts. Peel the shells off the walnuts. Slice the walnuts and the citron pieces.

7 Pour the citron syrup into the prepared ingredients from point 6 and mix well.
Roll the batter into 1cm balls and slightly squash them down.

8 Place the powdered red beans from point 5 into a big steamer in a square shape.
Put a scoop of the ingredients from point 2 onto these. Place the balls from point 7 on top of
and cover the ingredients from point 2 again. Steam them.

9 After 15 minutes, reduce the heat and steam them for another five minutes.
Place the cooked tteok in a dish and spread with the remaining red beans.
Cover them with the cloth and allow to cool.

Note

This traditional Korean rice cake was a must-have at the palace
on the kings' birthdays.
The cooking method is recorded in the 「Jeongrye euigwe」,
「Jinchan euigwe」 and in others. Duteop tteok is the
representative traditional Korean rice cake consumed in palaces
that seasoned rice powder with soy sauce. Its original name is
"Bonguri (summit) traditional Korean rice cake" and it is spelled
as Hubyeong (厚餅) in Chinese characters.

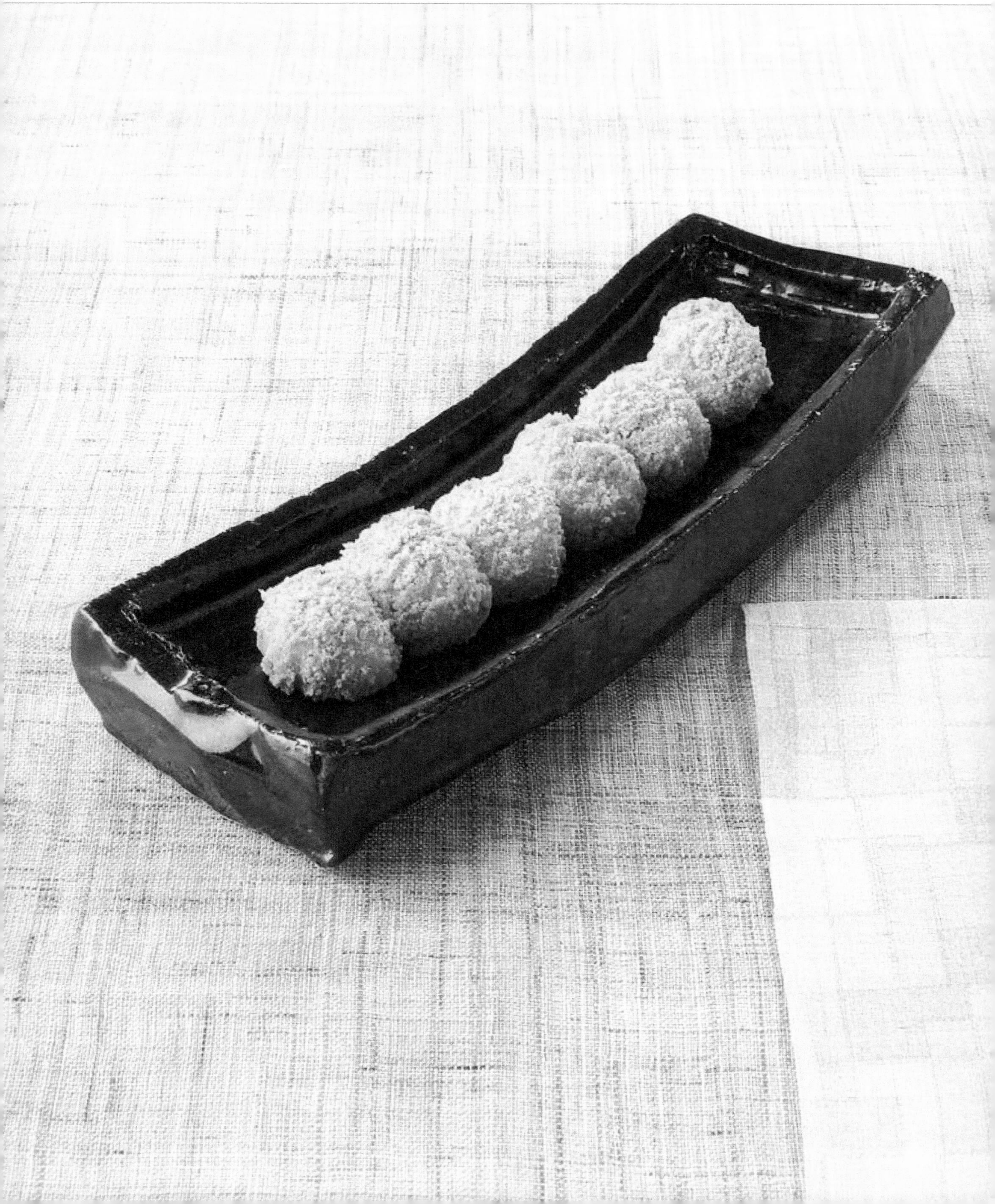

Bam danja
(Sweet Rice Balls with Chestnuts)

Ingredients

Glutinous rice 330g, chestnuts 160g, cinnamon powder 1/2 cup,
tangerine strips (tangerine marinated in honey) or citron cake 1 tablespoon,
honey 1 tablespoon, water, salt

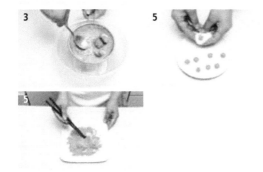

Cooking Directions

1. Soften the glutinous rice in water for no less than 2 hours.
   Sieve them to drain and remove the moisture and then grind the rice until it becomes like flour.
2. Lay a wet cotton cloth in a steamer, and place in the glutinous rice powder.
   Steam the flour thoroughly. Put it into a big bowl and beat it with a stick or spoon.
3. Add the chestnuts to some water and bring to the boil.
   Peel their shells once cool and sieve the flesh to make chestnut powder.
4. For the filling, finely chop the tangerine cake. Combine the chopped tangerine cake
   with 1/3 cup of chestnut powder, cinnamon powder and salt and mix well.
   Make pieces of batter which are 0.8cm in diameter.
5. Cut out pieces of the steamed glutinous rice from point 2 and make them into pieces of cake
   with a similar size of a chestnut. Spoon out a lump of the filling from point 4 and place it
   into the middle of the rice cake. Finally cover with honey and the remaining ground chestnuts.

Woomegi tteok
(Womegi Rice Cake)*

Ingredients

Glutinous rice powder 500g, non-glutinous rice powder 150g, crude rice wine (makgeolli) 1/2 cup, sugar 1/3 cup, water 2 tablespoons, salt 1/2 tablespoon, cooking oil 2 cups, a pinch of jujubes, For the syrup grain syrup 1 cup, water 100ml, ginger 10g

Cooking Directions

1 Mix the glutinous rice powder with the non-glutinous rice powder.
   Sieve and mix them well with the salt and sugar.
2 Add the crude rice wine into the mixed flour and stir well.
   Add some boiling water and beat for a long time to make dough.
3 Form the dough into balls 3cm in diameter and 1cm thick.
   Slightly press the top and the bottom of the balls.
4 Panfry the balls from point 3 at 180℃ in cooking oil until the color becomes a golden brown.
   (This is the woomegi)
5 Reduce the heat to 150℃ and cook the inside thoroughly.
6 For the syrup, combine the grain syrup, some water and ginger and bring to the boil.
7 Soak the cooked woomegi in the syrup from point 6 for a while and then place on a flat plate.
8 Garnish the woomegi with the small jujubes.

Note

Woomegi is a food that is made by covering traditional Korean rice cakes, fried in oil, with honey. It is easy to make and does not harden easily. This traditional Korean rice cake was made often, especially when the freshly harvested rice was produced. It is known to have been prepared for parties to the point that there is a saying that says, "There is no party without woomegi."
The dough should be lumped together. It looks good when it is shaped into a round shape and then a jujube is cut and added to make an engraving while pressing the center with the thumb. It does not harden easily for two to three days, and it tastes superb, thus it is ideal as a children's snack or desert. It is also called Gaeseong juak (Gaeseong-style rice cake).

---

* It is a type of fried cookies and works well as a dessert.

Maejakgwa
(Fried Ginger Cookies)

Ingredients

Flour 110g, salt 1/2 teaspoon, ginger juice 1 tablespoon, water 3-4 tablespoons, starch powder, cooking oil 3 cups, ground pine nuts 1 tablespoon

For the syrup sugar 150g, water 200ml, honey 2 tablespoons, cinnamon powder 1/2 teaspoon

Cooking Directions

1 Add the salt to the flour and sieve them.
Then add the water and the ginger juice and mix the flour to make dough.
2 Sprinkle the starch powder on a kitchen board. Put the dough from point 1 onto the starch powder.
Roll the dough out thinly and cut the dough into 5cm long and 2cm wide rectangular pieces.
Make 3 slits in the middle of each piece of dough.
3 Push each end of the dough through the center slit to make a ribbon shape.
4 For the syrup, add the sugar and water and bring them to the boil without stirring.
When the sugar has melted, add the honey and boil it over a low heat for about ten minutes.
Finally, add the cinnamon powder and mix well.
5 Increase the temperature of the cooking oil to 160℃.
Fry the dough from point 3 until they are golden brown.
6 Dip the snacks from point 5 into the syrup (the maejakgwa are now complete).
7 Decorate the maejakgwa in a dish, and sprinkle with the ground pine nuts.

Note

Maejakgwa is an oil-and-honey cookie that is coated with honey after putting salt and ginger juice into flour, which is then mixed together, rolled thin, turned over, by marking it slightly with a knife, and then fried. It is also referred to as maejagwa, maejatgwa, maejapgwa, maeyeopgwa and taraegwa. It is called maejakgwa by using the terms of a Japanese apricot flower and a tree sparrow since the cookie resembles a tree sparrow sitting on a Japanese apricot tree.

Seoyeohyang byeong
(Yam Cookies with Honey)

Ingredients

Yams 300g, ground pine nuts 1/2 cup, glutinous rice powder 50g,
honey 1/2 cup, cooking oil

Cooking Directions

1 Wash the yams thoroughly. After eliminating the moisture, peel the skins and cut into thin pieces.

2 Steam the yam pieces in a steamer for about seven minutes over a high heat.

3 Marinate the steamed yam pieces in honey.

4 Coat the yams marinated in honey with the glutinous rice powder and fry in cooking oil at 170℃.

5 Top with ground pine nuts before the fried yams cool.

Mogwa cheonggwa hwachae
(Yellow Quince Punch)

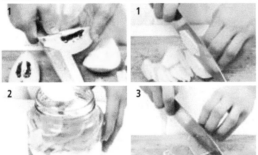

Ingredients

Yellow quinces 3kg, tangerines 180g, sugar 2 cups, pine nuts 2 tablespoons

Cooking Directions

1 Peel the yellow quinces and cut them into 1cm thick pieces.

2 Cut the tangerines into 0.5cm thick pieces (without peeling them).

3 Sprinkle the sugar onto the yellow quince and tangerine pieces. Pile them in a jar or a glass bottle on top of one another. Pour over plenty of sugar and seal it up tightly.

4 20 days later, put the marinated yellow quince and tangerine pieces in a bowl and add water. Add some pine nuts and allow them to float for decoration.

[Gangwon]

Gangwon is divided into the Yeongdong region and the Yeongseo region with the Taebaek mountain range serving as the demarcation line.

Yeongdong's seashore region is rich with different types of fishery products. Stored foods, such as salted seafood and sikhae (fermented fish) are well developed in their production here, and rice and condiments wrapped in seaweed, flakes of kelp oiled and toasted, and fish are also consumed often.

The Yeongseo region, with its many large mountains, used potatoes, corn, buckwheat, wheat, barley and other ingredients produced from the fields. Sometimes, main dishes were made by a mixture of potatoes, corn, millets and sweet potatoes instead of rice.

Gondalbi bap
(Seasoned Cirsium with Rice)

Ingredients

Rice 360g, Gondalbi (Gondeure, or cirsium) 300g, water 470ml, perilla oil 2 tablespoons, salt

Cooking Directions

1 Wash the rice thoroughly and soften it in water for 30 minutes.
2 Parboil the gondalbi, wash it in cold water and drain the water. Cut it into 3-5cm long pieces.
3 Season the parboiled gondalbi with the perilla oil and salt.
4 Cook the rice.
5 Place the seasoned gondalbi on top of the rice and cook for another few minutes.
   Mix well and serve the gondalbi rice.

Note

Gondalbi is a wild vegetable native to Mt. Taebaek, growing in the highlands about 70 m above sea level. It features a savory taste, plentiful nutrients, and a special fragrance. This vegetable has also been known to have helped the people who didn't have many foods to eat, as is known in the lyrics of the traditional Korean folk song 'Jeongseon Arirang.' It is gathered during the month of May.

Jogamja bap
(Rice with Millets and Potatoes)*

Ingredients

Gondalbi (Gondeure or cirsium) 200g, mackerel pikes 220g, onions 50g,
green hot peppers 30g, ginger 10g, water 400ml, red hot pepper paste 1/2 tablespoon, a pinch of salt
For the seasoning sauce soy sauce 2 tablespoons, crushed garlic 1 tablespoon,
chopped green onions 1 tablespoon, refined rice wine 1 tablespoon,
red hot pepper powder 1 tablespoon

Cooking Directions

1 Put a pinch of salt in boiling water and then put the gondalbi into the water.
Boil until the vegetable is thoroughly softened and then rinse in cold water.
2 Remove the head and innards of the mackerel pikes. Wash thoroughly.
3 Cut the onions and ginger into 0.3cm thick pieces. Slice the peppers diagonally into 0.3cm pieces.
4 Mix the ingredients for the seasoning sauce.
5 Spread the boiled gondalbi in a pan. Pour water over the gondalbi and add the red hot pepper paste,
mackerel pikes, onions, ginger and peppers.
6 Pour over the seasoning sauce and braise them.

---

* Potatoes and grains make the texture more rough and crunchy.

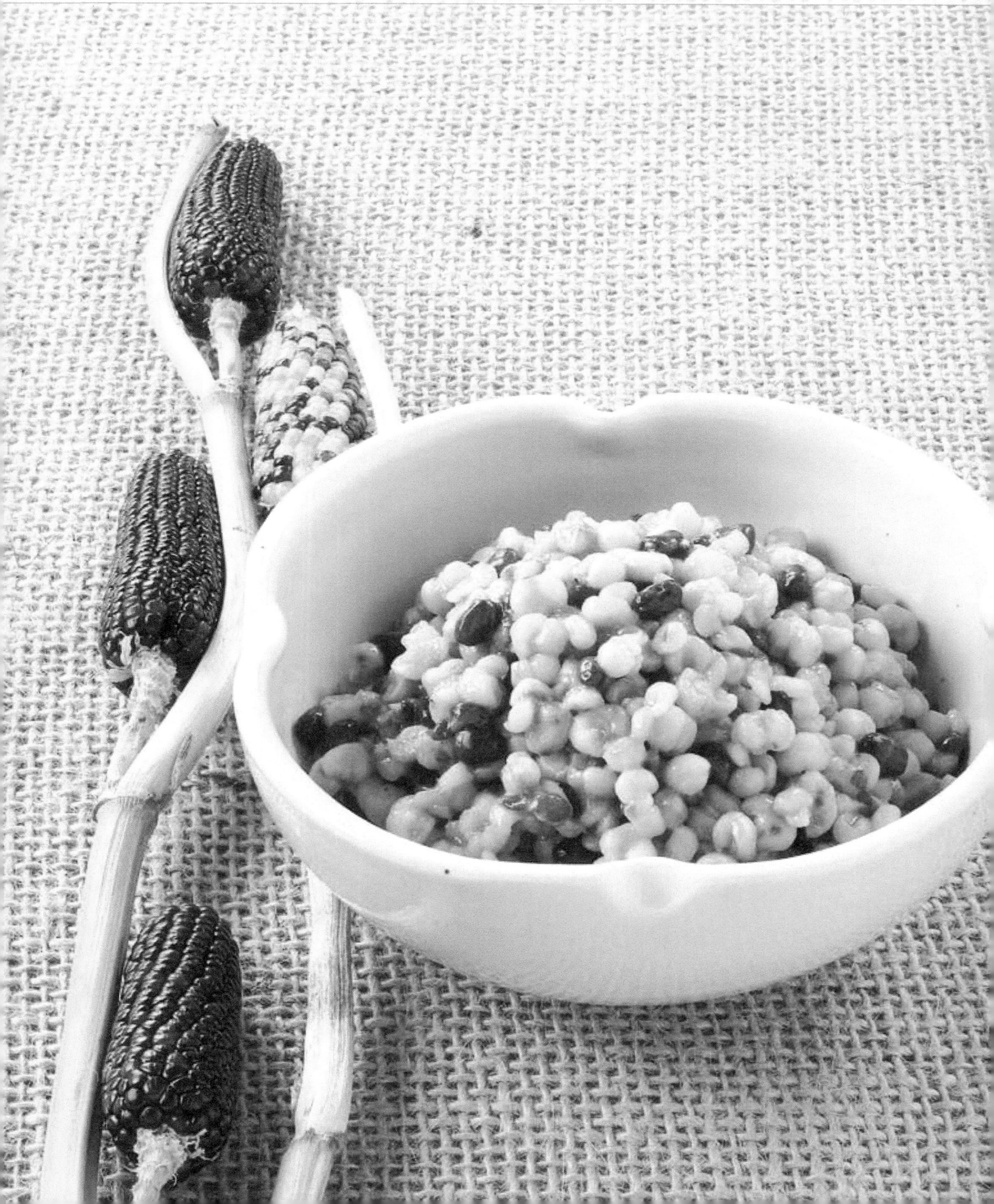

Chaloksusu neunggeun bap
(Corn Rice)

Ingredients

Neunggeun corns 290g, red beans 210g, water, sugar 1 cup, salt

Cooking Directions

1 Wash the corn and soften them in water overnight
2 Add water to the softened corn and red beans and bring them to the boil.
3 When the corn is cooked, season with salt and sugar.
  Stir them with a wooden scoop taking care not to burn over a low heat.

Note

'Neunggeun' refers to the pouring of water over corn and steaming them for the first time in order to peel off the husks.

Makguksu
(Buckwheat Noodles with Vegetables)

Ingredients

Buckwheat 2 1/2 cups, flour 160g, dongchimi broth (cold radish kimchi broth) 400ml,
1/2 head of kimchi, 1/2 radish of dongchimi, cucumbers 150g, eggs 50g, water for the dough 200ml,
crushed garlic 1 teaspoon, sesame oil 1 teaspoon, sesame salt 1 teaspoon, soy sauce, salt
For the chicken broth chicken 200g, radishes 100g, kelp 10g, onions 80g, ginger 10g,
green onions 10g, garlic 2, water 1L

Cooking Directions

1 To make the chicken broth, add all the ingredients in a pot and bring to the boil. Allow it to cool down and skim off the grease. Add the dongchimi broth and salt. Shred the chicken flesh into thick pieces and season with crushed garlic, sesame oil and sesame salt.

2 Mix the buckwheat and flour well. Mix the mixture with hot water and knead to make the dough to make noodles with a noodle maker.

3 Chop the cucumbers into strips and salt them. Squeeze the excess water out. Cut the dongchimi radishes into thin pieces and chop the kimchi into 1cm long pieces.

4 Parboil the buckwheat noodle in boiling water. Rinse them in cold water and shake off the excess water.

5 Place the parboiled noodles in a bowl and garnish them with the prepared cucumber strips, chopped kimchi and dongchimi radish, and egg strips. Pour the cooled broth into bowls and season with soy sauce and salt.

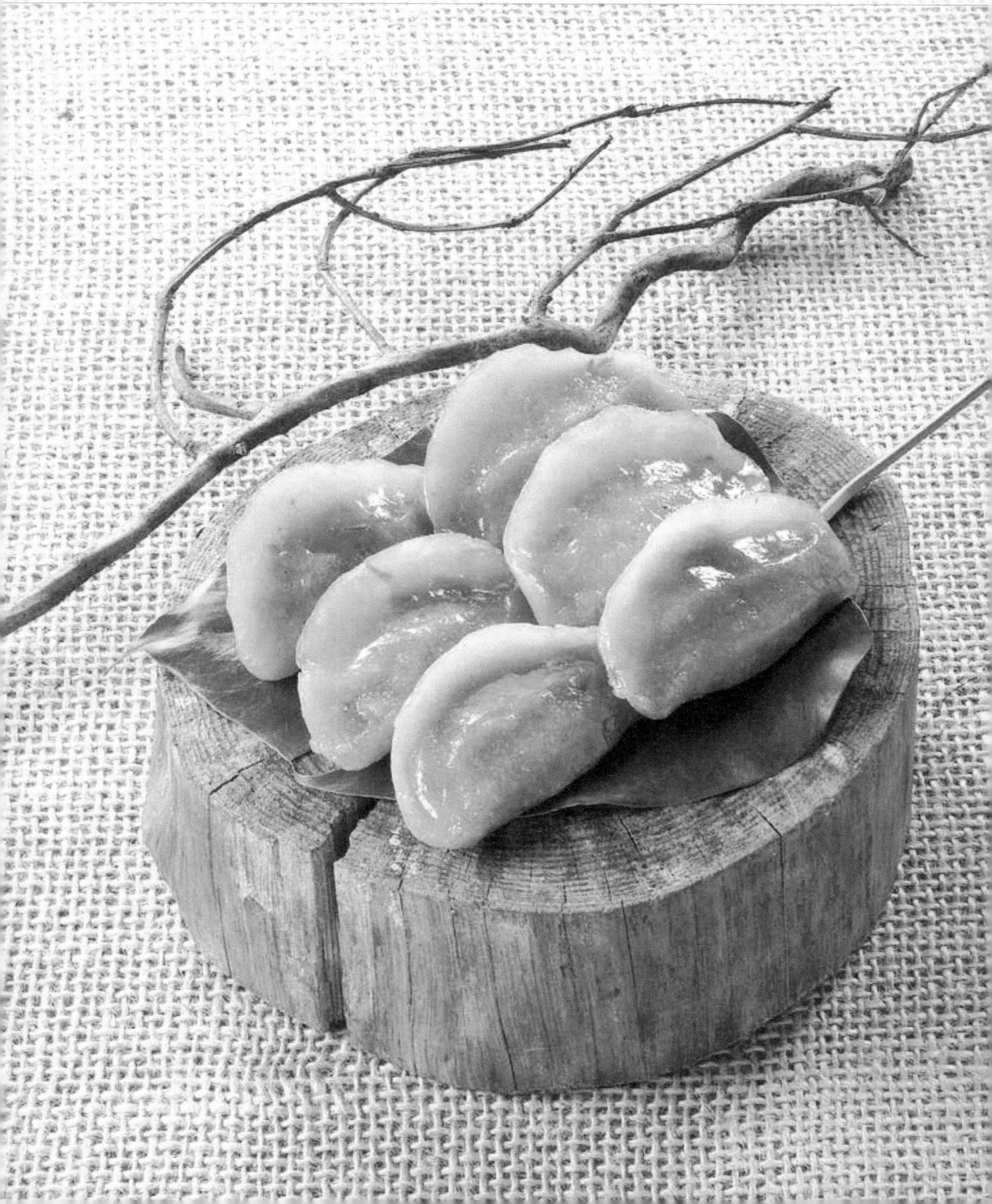

Chaemandu
(Buckwheat Dumpling)

Ingredients

Buckwheat flour (or starch) 3 cups, water 150ml, gat kimchi (mustard leaf kimchi) 200g, muk-namul (dried vegetables) 200g, water for dough 150 ml, perilla oil

For the seasoning chopped green onions, crushed garlic, salt, sesame oil, ground black pepper

Cooking Directions

1 Mix the buckwheat flour (or starch) with hot water.

2 Chop the gat kimchi into 0.5cm long pieces. Boil the dried vegetables and finely chop it up.
  Add the seasonings and mix well to make the filling for the dumplings.

3 Roll the dough from point 1 into a small ball.
  Stuff in the filling and make the dumplings by sealing the dumpling with the filling inside.

4 Place the dumplings in a steamer and steam for 20 minutes. Apply perilla oil on the steamed dumplings.

Note

Wash the gat kimchi with water and roughly cut it. Serve it with the dumplings.

Ojingeo bulgogi
(Grilled Squids)*

Ingredients

Squids 700g

For the seasoning sauce soy sauce 3 tablespoons, sugar 1 tablespoon,
chopped green onions 1 tablespoon, crushed garlic 1 tablespoon

Cooking Directions

1 For the seasoning sauce, combine the soy sauce, sugar, chopped green onions,
and crushed garlic and mix them well.
2 Remove the innards and legs of the squids. Peel off the skin. Spread it and make slits off 1cm across.
Marinate the prepared squids in the seasoning sauce.
3 Grill the seasoned squids.
4 Cut the grilled squids into 2cm long pieces.

Note

To prevent the squids from sticking to the grill, apply some
vinegar first to the grill. You can also add red hot pepper paste
to the seasoning sauce. Use a half-dried squids for a special
taste.

* For the strong marine products such as squids, it is better to season them like one would with
bulgogi.

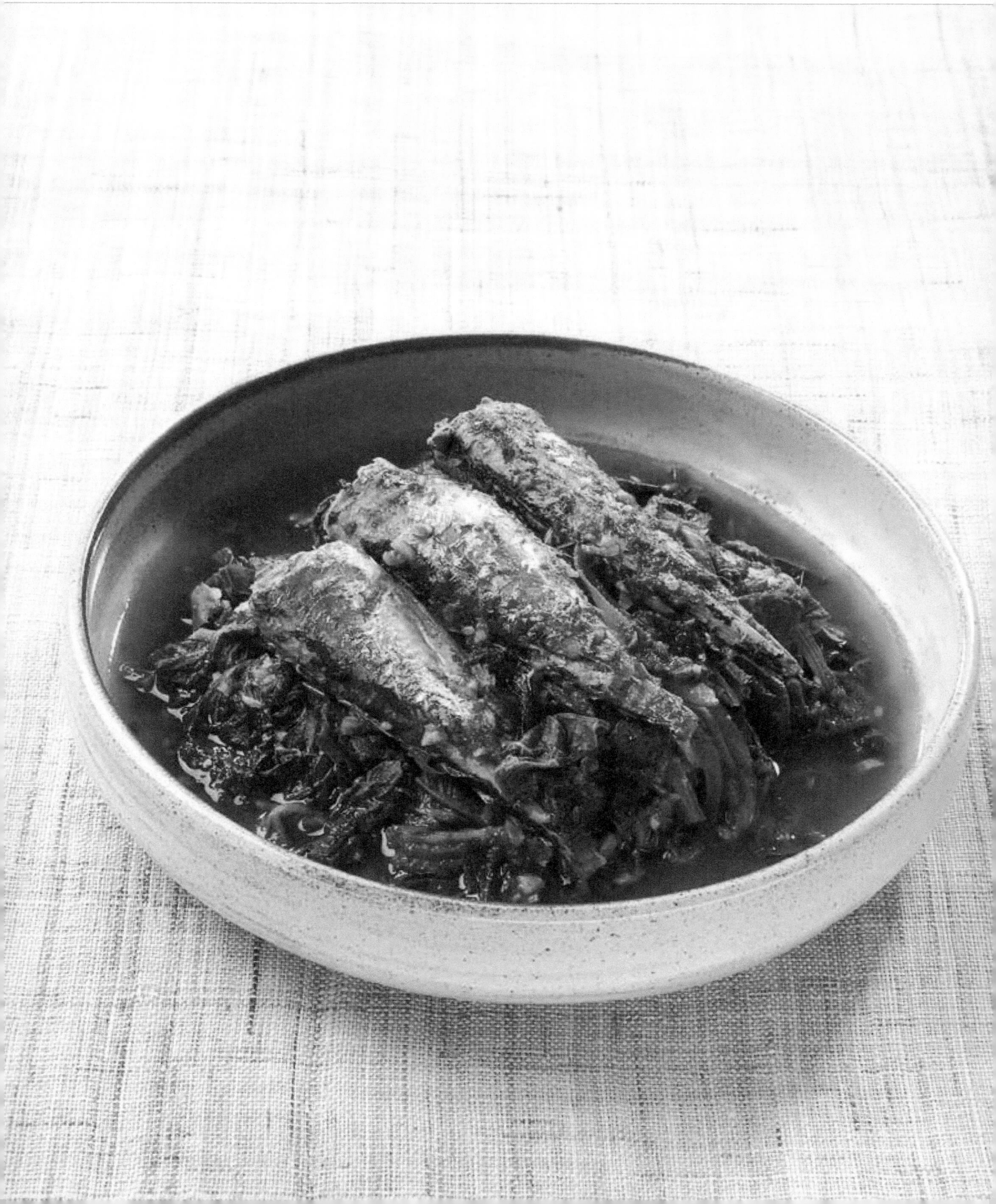

Gondalbi kkongchi jorim
(Braised Gondalbi and Mackerel Pike)

Ingredients

Gondalbi (Gondeure namul or cirsium) 200g, mackerel pikes 220g, onions 50g,
green hot peppers 30g, ginger 10g, water 400ml, red hot pepper paste 1/2 tablespoon, a pinch of salt
For the seasoning sauce soy sauce 2 tablespoons, crushed garlic 1 tablespoon,
chopped green onions 1 tablespoon, refined rice wine 1 tablespoon,
red hot pepper powder 1 tablespoon

Cooking Directions

1 Put a pinch of salt in boiling water and then put the gondalbi into the water.
   Boil until the vegetable is thoroughly softened and then rinse in cold water.
2 Remove the head and innards of the mackerel pikes. Wash thoroughly.
3 Cut the onions and ginger into 0.3cm thick pieces. Slice the peppers diagonally into 0.3cm pieces.
4 Mix the ingredients for the seasoning sauce.
5 Spread the boiled gondalbi in a pan. Pour water over the gondalbi and add the red hot pepper paste,
   mackerel pikes, onions, ginger and peppers.
6 Pour over the seasoning sauce and braise them.

Dak galbi
(Stir-fried Chicken Ribs)

Ingredients

Chicken 800g, cabbage 100g, sweet potatoes 50g, onions 50g, green onions 70g,
young green hot peppers 30g, 2 Chinese cabbage leaves, sesame leaves 10g, some lettuce,
Garae tteok (sticks of rounded rice cake), cooking oil
For the red hot pepper paste sauce red hot pepper paste 2 tablespoons, soy sauce 1 tablespoon,
red hot pepper powder 1 tablespoon, garlic 25g, ginger 10g, sugar 1 tablespoon,
sesame oil 1 teaspoon, refined rice wine 1 tablespoon, pears 50g, salt, a pinch of sesame seeds

Cooking Directions

1 Wash the chicken thoroughly and cut them into several pieces.
2 Grate the pears and finely chop the garlic and ginger.
  Mix the ingredients for the red hot pepper paste sauce.
3 Add the red hot pepper paste sauce to the chicken and mix well. Keep it as it is for 7 to 8 hours.
4 Roughly slice the cabbage, sweet potatoes, onions, green onions, young green hot peppers,
  and Chinese cabbage leaves (5×0.5×0.5cm).
5 Line a pan with cooking oil. Put in the vegetables, garae tteok (sticks of rounded rice cake),
  and chicken and stir-fry. When the chicken is cooked, cut it into bite sizes.
6 Clean the lettuce and sesame leaves thoroughly and serve with the chicken.

Note

There is a theory that traces back the Chuncheon-style 'dakgalbi' to the Shilla Dynasty era, about 1,400 years ago. The term, 'dakgalbi' was used in Hongcheon first. Hongcheon's dakgalbi is made by placing in broth and into a pot. To this day, this dish is consumed in Hongcheon and Taebaek. In Chuncheon, there was charcoaled dakgalbi which entailed the cooking of a chicken by placing it on a grill on top of the charcoal fire. The dish became Chuncheon dakgalbi as we know it today when the dakgalbi grilling plate emerged in 1971.

Gamja jeon
(Pan-fried Potato)

Ingredients

Potatoes 1kg, leeks 50g, small green onions 20g, red hot peppers 60g,
young green hot peppers 60g, a pinch of salt, cooking oil

Cooking Directions

1 Clean and peel the potatoes. Grate them. Put them to one side.
2 Cut the leeks and small green onions into 2cm long pieces.
 Finely chop the red hot peppers and young green hot peppers. Rinse them with water and deseed them.
3 Combine the grated potatoes and starch, leeks and small green onions and mix well.
 Season with salt.
4 Pour some oil into a heated pan and scoop the batter from point 3 and drop it into the pan
 to make the pancakes. Top the batter with the chopped red hot peppers and young green hot peppers.
 Pan-fry both sides until they are golden brown.

Ojingeo sundae
(Stuffed Squids)

Ingredients

Squids 1kg, glutinous rice 100g, starch 150g, eggs 250g, burdocks 70g, cucumbers 70g, carrots 70g, soy sauce 2 tablespoons, salt, sesame oil, some broth (anchovy, kelp, water)

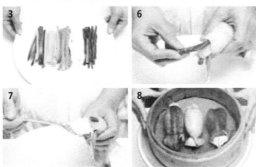

Cooking Directions

1 Remove the innards and bones of the squids by hand.
  Wash the body and season with salt. Remove the excess moisture.
2 Fry the egg white and yolk separately into very thin layers.
3 Cut the egg white, yolk, cucumbers, carrots, and burdocks into thick strips (6×0.5×0.5cm).
4 Salt the cucumbers and squeeze the excess water out. Stir-fry them slightly.
  Parboil and stir-fry the carrots slightly.
  Pour the anchovy broth and soy sauce mixture over the burdocks and braise them.
5 Soften the glutinous rice in water. Steam it in a steamer.
  Season the steamed glutinous rice with sesame oil and salt.
6 Put some starch powder into the inner body of the squids and then shake it off.
  Insert the egg strips, cucumbers, carrots, and burdocks into the squids' body.
7 Stuff the body with the steamed glutinous rice.
8 Skewer the stuffed squids crosswise and steam it in a steamer over a high heat for 15 minutes.
9 When the food has completely cooled down, cut it into bite sizes.

Note

Gangwon Province is famous for 'sundae' or 'Korean sausage' foods made of pork, squids, and Pollack.
You can also insert seasoned tofu and vegetables into the squids' body.

Memil chong tteok
(Buckwheat Pancakes)*

Ingredients

Buckwheat 2 cups, water 600ml, salt 1 teaspoon, some cooking oil
For the filling of the pancakes gat kimchi (mustard leaf kimchi) 300g,
chopped green onions 1 tablespoon, crushed garlic 1 tablespoon,
sesame oil 2 tablespoons, sesame salt 2 tea spoons

Cooking Directions

1 Season the buckwheat with salt. Pour the water and mix well.
2 Shake off the seasonings of the gat kimchi (mustard leaf kimchi).
  Wring out the water and chop the kimchi into small pieces.
3 Add the chopped green onions, crushed garlic, sesame oil, and sesame salt
  to the chopped gat kimchi and mix them well to make the filling.
4 Pour cooking oil in a pan and scoop the buckwheat batter and drop it into the pan.
  Spread the batter thinly.
5 When one side of the batter is cooked, pan-fry the other side and place the filling horizontally
  over 1/3 of the pancake. Pan-fry the pancake with the filling inside, rolling it from the front.

Note

Buckwheat was cultivated en-mass from the ancient times to the point that it was recorded as a hardy plants to relieve famine in the Guhwang Byeokgokbang , a book published during King Sejong's reign in the Joseon Dynasty period. Memil chong tteok is mentioned as 'Gyeonjeonbyeong' in the Yorok published in 1680. In the Jubangmun during the latter end of the 1600s, it was called the 'Gyeomjeolbyeongbeop.' The term, 'Chong tteok' was used for the first time in 1938 in Cooking of the Joseon Dynasty Period . Buckwheat, which is the main ingredient for the memil chong tteok, is one of the representative crops from Gangwon. The buckwheat harvested from the gravel area at a high altitude place was particularly well renowned for its quality. Buckwheat grows mostly in Gangwon and Gyeongbuk and it is similar to the 'bing tteok' of Jeju Island.

As for the ingredients for the pancake fillings, neunjaengi namul (seasoned vegetables) or dried pepper leaves were used after they were soaked in water. These days, kimchi and pork are mixed in together.

---

* Chong tteok is well received in the West, as they are similar to pancakes or crepes.

Gangneung sanja
(Fried Dough of Sweet Rice Rolled in Popped Rice)

Ingredients

Glutinous rice 720g, liquor 2/3 cup, grain syrup 1 1/2 cups, cooking oil

Cooking Directions

1 Wash the glutinous rice thoroughly and soften it in water. Grind finely and sift it.
  (Soften for 7 days in summer and 14-15 days in winter.)
2 Mix the glutinous rice powder with the liquor. Lay a wet cotton cloth in a steamer.
  Put the dough on the cloth and steam. Strongly pound and beat the dough in a mortar.
3 Sprinkle flour on a chopping board and roll the dough into a thin layer.
  Cut it into some pieces. Spread the dough pieces on a warm surface and allow it to dry thoroughly.
  Don't let any wind blow in.
4 Stir-fry the glutinous rice in a pot to make maewha (popped rice).
5 When the dough is completely dry, put it in some oil over a low temperature at first and then steadily
  increase the heat. Coat the fried food with the grain syrup and the maehwa (popped rice)

Note

'Maehwa' refers to the popped rice and 'Maehwa sanja' refers
to the food coated with maehwa.

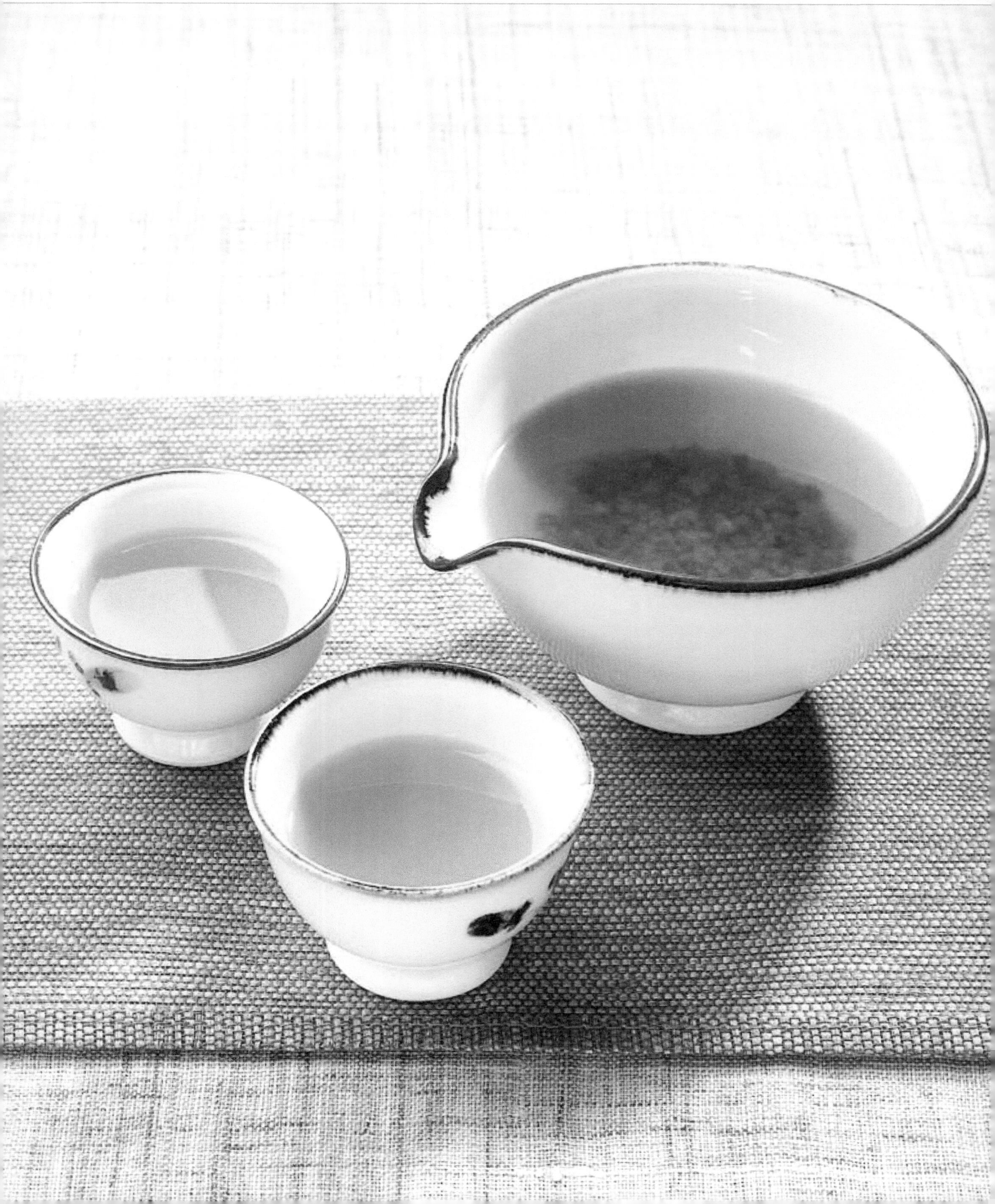

Memil cha
(Buckwheat Tea)

Ingredients

Buckwheat 1 cup, water 2L

Cooking Directions

1 Remove the husks from the buckwheat.
2 Pour water over the buckwheat and cook it like you would rice. Dry and stir-fry the cooked buckwheat.
3 Put the stir-fried buckwheat in a pot, pour in some water and bring it to the boil.

Songhwa milsu
(Pine Pollen Flour Punch)

Ingredients

Pine pollen flour 1 tablespoon, water 200ml, honey 3-5 tablespoons, pine nuts 5

Cooking Directions

1 Boil and then cool down the water. Add the honey to the water and ensure it melts well.
2 Add the pine pollen flour.
3 Serve the punch with floating pine nuts.

Hobak sujeonggwa
(Pumpkin Punch)

Ingredients

Mature pumpkin 3kg, cinnamon 50g, ginger 50g, some pinches of dried persimmons, pine nuts and walnuts, yellow sugar 200g, some water

Cooking Directions

1. Peel the ginger and wash it thoroughly. Slice it into thin layers.
   Pour water over the ginger and bring it to the boil.
2. Wash the cinnamon, pour water over the cinnamon and bring it to the boil.
3. Peel the mature pumpkin. Remove the seeds and empty the inside.
   Cut into some pieces and put them into a pot with water. Bring to the boil.
4. Add the ginger water from point 1 and the cinnamon water from point 2 to the pumpkin from point 3.
   Bring them to the boil.
5. Sieve the water from point 4 through a clean cotton cloth.
   Add the yellow sugar and bring them to the boil again. (Punch completed)
6. Wipe the surface of the dried persimmons with a wet cotton cloth.
   Take off their peaks and cut them lengthwise into two pieces. Deseed them.
7. Parboil the walnuts in boiling water to remove their inner shell.
8. Insert the walnuts into the dried persimmons. Press them with a suitable tool (or just use your fingers) to make dried persimmon wraps. Cut the wrap into 0.5cm long pieces.
9. Pour the cooled punch into a bowl. Sprinkle with the dried persimmon wraps and pine nuts.

[Chungbuk]

Chungcheongdo is the only inland area of Korea that does not come into contact with the ocean since it is located in the very center of the Korean peninsula. Because there are many hilly mountainous areas and wide plains, rice farming was advanced here and large amounts of grains such as rice, barley and beans and sweet potato, along with peppers, Chinese cabbage, mushrooms and many others are still produced.

Instead of the ocean, interaction with the region's interior waters developed, and thus dishes using fresh water fish, such as Far Eastern catfish, Japanese eel, crucian carp, freshwater mandarin fish and others, became highly developed. People of this region do not use seasoning often but prefer dishes offering the inherent taste of the ingredients, which leaves a humble but savory flavor.

Traditional kimchi here included using extensive amounts of garlic and peppers. Salt was often used instead of salted seafood and thus the kimchi was referred to as 'jjanji' (salty kimchi).

During the winter, Chinese cabbage jjanji was made while young radish jjanji was made during the summer. Kimchi from this area is also characterized by its lack of broth. This region is famous for 'gat jjanji,' which is made from mustard leaves. They would cut the mustard leaves into strips, mix them with vinegar, salt, sugar and sesame oil, place in a jar and then would eat them after one night.

This region is also renowned for haejangguk (hot pot for hangovers) made with Chinese cabbage, seonji (clotted ox blood) or tripe.

Due to extensive bean production, people here ground beans into powder to use for steamed vegetable dishes or to make flour dough, or porridge.

Hobak cheong
(Pumpkin Porridge)

Ingredients

Mature pumpkin 1.5kg, chestnuts 200g, jujubes 300g, gingkoes 20g, ginger 50g, ginseng roots 2, honey 200g, glutinous rice powder 100g, water 3 tablespoons

Cooking Directions

1. Cut the top of the mature pumpkin out to the size of your palm to make a lid. Remove the seeds and empty the inside.
2. Peel the chestnuts. Pan-fry the gingkoes in oil and remove the inner shells. Peel the ginger and chop up it into thin pieces.
3. Mix the glutinous rice powder with hot water and make dough to make small round dumplings
4. Put the prepared chestnuts, jujubes, gingkoes, ginger, ginseng, and dumplings in the pumpkin and pour honey over them. Close the lid and then thoroughly steam in a steamer.

Note

Pumpkin has been known to be beneficial after one is pregnant. Koreans enjoy pumpkin juice or pumpkin porridge.

Okgye baeksuk
(Stuffed Chicken Soup)

Ingredients

Okgye (a type of chicken) 1kg, glutinous rice 335g, chestnuts 100g, jujubes 10g,
ginseng roots 2, hwanggi (a herb) 4, Yulmu (Job's Tears) powder 3 tablespoons, hand-rolled noodles,
water, green onions 35g, garlic 20g, a pinch of sesame seeds, black pepper, and salt

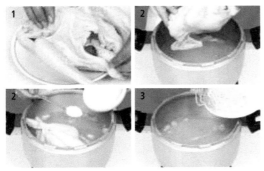

Cooking Directions

1 Remove the innards of the chicken, wash the body well.
  Stuff its insides with the jujubes, chestnuts, glutinous rice, and ginseng roots.
2 Place the stuffed chicken in a pressure cooker and pour water over it.
  Add the garlic and bring all the ingredients to the boil. When they are half-cooked,
  add the hwanggi and Job's tears powder and cook them thoroughly.
3 Take out the cooked chicken and place it into a bowl.
  Add the chopped green onions and the hand-rolled noodles to the broth in the pressure cooker.
  Boil them thoroughly and season with sesame salt, salt, and black pepper.

Note

Okgye is a chicken native to the Okcheon region, which is
characterized by its black legs. The people living in this region
used to eat this chicken by cooking it with various traditional
oriental herbs to remove any smell from the meat, adding
noodles or glutinous rice into the broth.

Kong guk
(Chilled White Bean Soup)

Ingredients

Soybean thin porridge 5 cups, tofu 250g, bean sprouts 200g, carrots 140g, potatoes 300g, green onions 10g, garlic 10g, red hot pepper powder 1/2 tablespoon, a pinch of salt

Cooking Directions

1 Put the bean sprouts in a pan and pour water over them. Slightly parboil the bean sprouts.
2 Cut the carrots and potatoes into rectangular pieces (3×1×0.3cm). Add some salt and slightly parboil them.
   Cut the tofu into pieces of the same size as the carrots and potatoes.
3 Place the slightly-cooked bean sprouts, carrots and potatoes into a pot
   and pour the soybean porridge over them. Bring them to the boil.
4 When the soup begins to boil, add the tofu, crushed garlic and chopped green onions.
   Skim off the foam while the soup is boiling. Season it with red hot pepper powder and salt.

Kongbiji tang
(Pureed Soybean Soup)

Ingredients

Soybeans 160g, pork ribs 100g, kimchi 50g, radishes 20g water 800ml,
salted shrimps 2 tablespoons, salt 1 teaspoon
For the seasoning for the pork ribs soy sauce 1 tablespoon, chopped green onions 2 tablespoons, crushed garlic 1 tablespoon, crushed ginger 2 teaspoons, sesame oil 1 teaspoon, a pinch of black pepper

Cooking Directions

1 Wash the soybeans and soften them in water overnight.
  Remove their husks by rubbing them by hand. Sieve them to remove the excess moisture.
  Put the soybeans and the same quantity of water into a millstone or a grinder to finely grind them.
2 Cut the pork ribs into 3cm long pieces. Cut some slits on them. Mix the pork with the seasonings.
3 Wash and cut the radishes into thick pieces (5×0.5×0.5cm) and cut the kimchi into 2cm wide pieces.
4 Line a thick pan with some cooking oil. Stir-fry the pork ribs. When the pork ribs are half-cooked, add the radishes and kimchi and stir-fry them thoroughly.
5 Pour the soybean porridge, made in point 1, over the ingredients from point 4.
  Simmer them over a medium heat. When the porridge becomes a clear color, season it with salt and salted shrimp.

Deodeok gui
(Grilled Deodeok Roots)

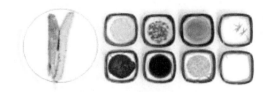

Ingredients

Deodeok roots (roots of the Codonopsis lanceolata) 300g, some vinegar
For the seasoning red hot pepper paste 2 tablespoons, soy sauce 2 tablespoons,
sugar 2 tablespoons, chopped green onions 2 teaspoons, crushed garlic 1 teaspoon,
sesame salt 1 teaspoon, sesame oil 1 teaspoon
For the serum sesame oil 1 tablespoon, soy sauce 1 tablespoon, deodeok 240g, red hot pepper paste

Cooking Directions

1 After peeling the deodeok roots, wash them thoroughly.
2 Cut the roots lengthwise into 2 pieces and roll them out flat.
3 Make seasoning by mixing the seasoning ingredients mentioned above.
4 Apply the serum and then the vinegar to the deodeok roots.
  Slightly grill the deodeok roots on a grilling instrument.
5 Grill them while applying the seasoning.

Note

The wild deodeok is a specialty of Suanbo, which located
around the Woraksan National Park. This appetite stimulating
food has been called 'Sasam' or 'Baeksam,' which is a type of
ginseng. Deodeok gui is very popular among tourists as well as
the local residents.

Doribaengbaengi
(Pan-fried Fresh-water Fishes)

Ingredients

Fresh-water fishes (smelts, daces, etc.) 170g, fresh ginseng 10g, carrots 10g,
green onions 10g, young green hot peppers 15g, red hot peppers 15g
For the seasoning sauce red hot pepper paste 3 tablespoons, crushed garlic 1/2 tablespoon, crushed
ginger 1/2 tablespoon, sugar 1/2 tablespoon, water 3 tablespoons

Cooking Directions

1 Clean the fishes. Place them in a pan in a circular manner.
Pour over some cooking oil and fry until they turn a beautiful yellow.
2 Slice the carrots and green onions into strips of 5×0.2×0.2cm.
Slice the fresh ginseng and red hot pepper diagonally into 0.3cm pieces
3 Mix the ingredients to make the seasoning sauce.
4 When the fish is fried, pour out the oil and cover the fish with the seasoning sauce.
Garnish with the vegetables from point 2 and cook slightly.

Note

Doribaengbaengi that consolidated its position as the
representative local food of the area near Euirimji (lake) in
Jecheon and Daecheong Dam used to refer to the dish of small
fresh water fishes that are rolled around in a pan. According
to the people in Joryeong-ri, an elderly man from northern
Korea started selling this under the name of braised fishes.
Since then, it was called by many names, such as fried fish or
braised daces, but a customer one day said, "Please give me
dori baengbaengi, which is placed in a pan in a round, rolled
shape," which is how this name came about.

Dotori jeon
(Pan-fried Acorn)

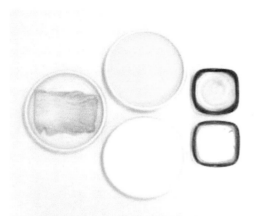

Ingredients

Acorn powder 150g, flour 110g, washed kimchi, cooking oil, water 600L, salt 1 teaspoon

Cooking Directions

1 Mix the acorn powder, flour, and salt. Sieve the mixture.
2 Pour the water over the mixture from point 1. Mix well.
3 Line a pan with cooking oil. Place in a leaf of kimchi. Spread the mixture from point 2 over the pan.
4 Fry both sides.

Chick jeon
(Pan-fried Arrowroot)

Ingredients

Arrowroot starch 160g, flour 55g, young green hot peppers 20g,
red hot peppers 20g, young pumpkins 80g, water 400ml, a pinch of salt, oil

Cooking Directions

1 Add the flour and water into the arrowroot starch and mix them well. Sieve the mixture.
2 Cut the young pumpkins into thick pieces (5×0.3×0.3cm).
  Slice the red hot peppers, and young green hot peppers diagonally into 0.3cm pieces.
  Mix them with the mixture from point 1.
3 Line a heated pan with cooking oil. Spread the mixture over the pan and fry.

Note

Seasoning soy sauce (soy sauce, sesame oil, sesame seeds,
chopped green onions, and crushed garlic) also goes well with
chicken jeon.

Pyogo jangajji
(Pickled Shiitake Mushrooms)

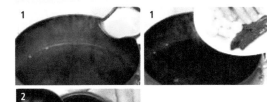

Ingredients 1

Dried shiitake mushrooms 100g, dried hot peppers, garlic 30g, water, soy sauce 4 cups, ginger juice 1 tablespoon, salt 1 tablespoon

Ingredients 2

Dried shiitake mushrooms 100g, soy sauce 2 cups, light soy sauce 2 cups, dextrose syrup 2 1/2 cups, sugar 2 cups
For the kelp broth kelp 20g, garlic 30g, ginger 20g, onions 70g, dried hot peppers 5, water 7L

Cooking Directions 1

1 Combine the soy sauce, water, ginger juice, salt, garlic and dried hot peppers and mix them well.
Bring the mixture to the boil and then allow to cool.

2 Place the dried shiitake mushrooms in a jar. Pour the soy sauce mixture made from point 1
over the mushrooms.

Cooking Directions 2

1 Soften the dried shiitake mushrooms in water. Take off the tips and remove the excess moisture.

2 Pour the water into a pot and add the kelp, garlic, ginger, onions and dried hot peppers.
Bring them to the boil for 20 minutes.
Sieve them through a clean cotton cloth to gain a clear broth.

3 Add the soy sauce, light soy sauce, sugar, and dextrose syrup to the broth from point 2
and boil them down until the broth is reduced by two thirds.

4 Put the softened shiitake mushrooms into the broth from point 3.
Take out the solid ingredients from the broth. Boil the remaining broth for another 5 minutes
and then allow to cool.

5 Place the cooked shiitake mushrooms in a jar and pour the soy sauce from point 4
over the mushrooms.

[Chungnam]

In Chungnam, an ample amount of grain is produced as it has extensive farming lands such as the Yedang plain and the Geumgang drainage plain. The region is also rich in fishery products since it borders the Western seashore. Its cuisine is characterized by a more natural taste, since it does not use much seasoning, just like the cuisine from Chungbuk.

Meanwhile, the amount of food produced from this region is ample. Doenjang (soybean paste), cheonggukjang jjigae (thick soybean paste stew), porridge, wheat noodle, sujebi (dough flakes soup), and others are consumed along with barley rice. In the summer chicken is served and oysters or sea shells are served in the winter to cook tteokguk (rice cake soup) and kalguksu (hand-rolled noodle soup). In addition, pumpkin porridge, pumpkins goji tteok (cake made with dried pumpkin slices) and pumpkin kimchi, which used old pumpkins, were consumed significantly.

Ogolgye tang
(Chicken Soup with Black Bone)

Ingredients

Ogolgye chicken 1,

various traditional Korean medicinal herbs

eomnamu (Kalopanax pictus), cheongung (cnidium rhizome), danggui(Chinese angelica), hwanggi (a herb), gugija (Chinese matrimony vine), changchul(Altractylodis Rhizoma), gamcho (Glycyrrhiza glabra), deer horns, jujubes, chestnuts, water 3L, salt 2 tablespoons

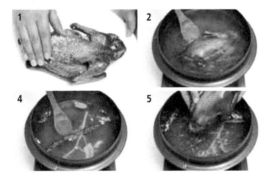

Cooking Directions

1 Remove the innards of the chicken and rinse thoroughly. Rub the chicken with salt.

2 Parboil the prepared chicken in boiling water.

3 Peel both the outer and the inner shells of the chestnuts. Wash the eomnamu, cheongung, danggui, hwanggi, gugija (Chinese matrimony vine), changchul, gamcho, deer horns, and jujubes thoroughly.

4 Put the Korean medicinal herb ingredients in a pot and pour water over them.
Boil them thoroughly until the soup has a rich fragrance.

5 Add the jujubes, chestnuts and chicken into the broth and bring them to the boil again.
The soup is now ready.

Jeoneo gui
(Grilled Gizzard Shad)

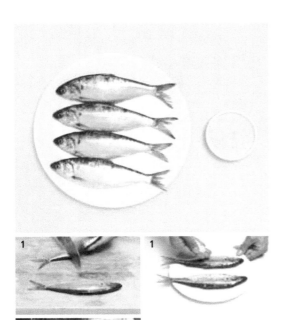

Ingredients

Gizzard shads 3, salt 1/2 tablespoon

Cooking Directions

1 Remove the scales of the gizzard shads.
  Wash them thoroughly and remove the excess moisture. Salt them.
2 Evenly grill both sides of the fish, frequently turning them over until they turn golden brown.

Hobakgoji jeok
(Pan-fried Pumpkin Slices)

Ingredients

Dried pumpkin slices 100g, small green onions 100g, beef 200g,
glutinous rice powder 100g, cooking oil 1 tablespoon, water 100ml
For the seasoning for the beef soy sauce 1 tablespoon, chopped green onions 2 teaspoons,
sugar 1 tablespoon, sesame oil 1 teaspoon, sesame salt 2 teaspoons, black pepper 1/3 teaspoon
For the seasoning for the dried pumpkin slices chopped green onions 2 teaspoons,
soy sauce 1 tablespoon, sesame oil 1 teaspoon, sesame salt 2 teaspoons

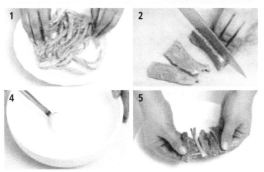

Cooking Directions

1 Select the thick dried pumpkin slices. Soften them in water. Season them with the seasonings.
2 Cut the beef into thin pieces (6×1.5×0.5cm) and season them with the seasonings.
3 Cut the small green onions into 6cm pieces. Season them with sesame oil.
4 Mix the glutinous rice powder and water well.
5 Skewer the dried pumpkin slices, beef and small green onions alternately to make brochettes.
  To make them look well presented place the pumpkin slices on both ends.
6 Soak the brochettes in the mixture from point 4.
  Line a heated pan with cooking oil and fry the brochettes.

Kkotge jjim
(Steamed Blue Crab)*

Ingredients

Blue crabs 1kg, soybean paste 1 tablespoon, some water

For the horseradish soy sauce soy sauce 4 tablespoons, horseradish powder, some water

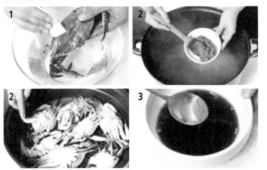

Cooking Directions

1 Choose blue crabs caught during April or May. Clean them in running water with a brush.

2 Pour water into a pot and mix the soybean paste with water.
  Lay the blue crabs on their backs in the water. Steam thoroughly.

3 Mix the horseradish powder and water well. Add it into the soy sauce and mix well.

4 Serve the crabs with the horseradish soy sauce.

---

* Steamed crab is also popular on the eastern coast of the US, the UK and Australia.

Seodae jjim
(Steamed Tonguefish)

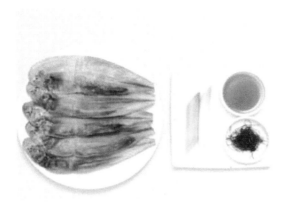

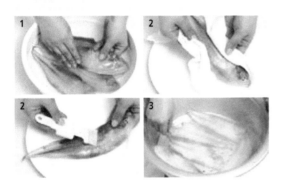

Ingredients

Dried tonguefish 100g, green onions 10g, sesame oil 1 tablespoon, a pinch of sliced red peppers

Cooking Directions

1 Wash the dried and salted tonguefish with water and remove the excess moisture with a clean cotton cloth.

2 Brush them with sesame oil.

3 Put the tonguefish into a steamer and steam thoroughly.

4 Add the chopped green onions and sliced red peppers on top of the tonguefish. Cover and steam again. When the steam rises up again, turn off the heat.

Note

Bakdae is another name for Seodae(tonguefish), which is used in Chungcheong Province. The flat shape looks like a tree leaf and the sole of a shoe. It is used for diverse dishes since the meat tastes excellent. It is usually salted and dried like a flatfish. This fish is especially popular in the Seocheon area. They enjoy the dried ones by steaming them or frying them in oil.

Hodu jangajji
(Pickled Walnuts)

Ingredients

Walnut flesh 240g, beef 100g, water 140ml,
soy sauce 3 tablespoons, dextrose syrup 1 tablespoon
For the seasoning for the beef soy sauce 1 teaspoon, chopped green onions 1 teaspoon, crushed garlic
1/2 teaspoon, sesame salt 1 teaspoon, sesame oil 1 teaspoon

Cooking Directions

1 Put the walnuts in some boiling water.
   When the walnuts begin to float to the surface turn off the heat and let them cool
   for 10 minutes to remove the tannin. Rinse in cold water and place in a tray.
2 Finely mince the beef and season with the ingredients.
   Take the minced and form balls which are 1.5 cm- 2cm in diameter.
3 Mix the soy sauce and water and bring to the boil.
   Add the beef balls from point 2 and the walnuts from point 1 and braise them.
4 When the beef balls and walnuts are braised remove them
   and over pour the dextrose syrup and mix well.

Insam jeonggwa
(Candied Ginseng)

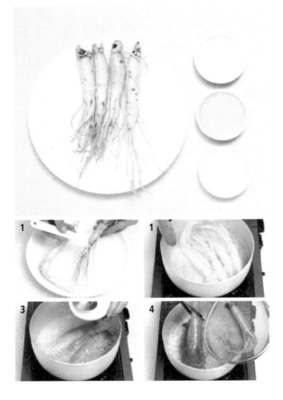

Ingredients

Ginseng (fresh) roots 4, sugar 6 tablespoons, dextrose syrup 2 tablespoons, honey 1 teaspoon, water

Cooking Directions

1. Wash the ginseng with a brush in running water and then add to a pot and boil thoroughly.
2. Put the boiled ginseng in another pot and pour over the boiled water from point 1.
   Add sugar (ginseng: sugar = 2:1) and simmer gradually bringing to the boil.
   Be careful not to stir while the sugared water is boiling.
3. When the sugared water is reduced by half, add the starch syrup and keep it simmering.
   Make sure you do not stir.
4. When the sugared water is almost completely boiled down and the color of the ginseng
   turns a transparent red and is shiny, pour in the honey and stir.

Note

The ginseng jjeonggwa has especially been popular among
men because it is known to increase their stamina.

Bori sikhye
(Fermented Barley Rice Punch)

Ingredients

Boiled barley rice 630g (for three persons), powdered malt 120g, water 3L, malt 1/2 cup, sugar 2 cups

Cooking Directions

1 Mix the powdered malt with the water by strongly rubbing the powder through your fingers to make the malt water.
2 Leave the malt water as it is for a while then pour away only the top of the water leaving the malt silt at the bottom.
3 Put the malt into the boiled barley rice and beat together.
   Pour the malt water from point 2 over the boiled barley rice mixed with the malt.
4 Allow it to ferment overnight at a temperature of 50-60℃.
   When the barely rice grains float to the surface, sieve the grains from the malt water and put them aside.
   Add sugar to the malt water and bring to the boil and then allow to cool.
5 Serve with the grains.

Note

Sikhye (rice punch) is fermented rice with malt. It is sometimes consumed with pine nuts floating in the punch. Beverages consumed without rice grains are called gamju, or simply a sweet rice drink. In the book, 『Making Korean Food』, it is said that it is better to make "Sikhye with non-glutinous rice than glutinous rice sin ce non-glutinous rice becomes softer." In the beginning, honey was added too but after the 『Cooking of the Joseon Dynasty Period』 was written, sugar has mostly been used. Moreover, citrons, pomegranates, jujubes and chestnuts are used to add color and taste to sikhye. It is mentioned in the 『Sumunsaseol』 that, "when citrons that are not peeled are placed in the rice while it is cooking, the taste becomes refreshing and the rice grains remain intact while the color turns white and the taste becomes sweet."

Jeonbuk

Jeonbuk includes the Honam plain, the largest area of farming land in Korea, borders the Yellow Sea and has several mountains. Thus, the area is considered the center of the farming culture of Korea, so much so as to the point that 16% of Korean rice is produced in this area. Moreover, the area is famous for the production of ginseng, bellflowers and schizandra from the mountainous areas as well as for its fishing industry. Jeonju is famous for rice: sweet potato rice; nonglutinous millet rice; bibimbap (cooked rice served with vegetables and meat to mix in the bowl at the table); as well as bean sprout soup and bean sprout rice. As for the broth of the tteok guk (rice cake soup), consumed during the New Year, anchovies, beef or peacock meat is used. Bean sprout soup or miyeok guk (seaweed soup) is seasoned only with salt.

Kimchi here does not take any pepper powder except for the gimjang period (kimchi-making season). Kimchi here is made as follows: Salt down some Chinese cabbage for several hours and then dry them. Grind glutinous rice paste, rice, garlic, ginger, soaked peppers, and red peppers and put salted seafood or salt into the ground mixture. Then put all these together with the radish strips and combine them well to make the kimchi ingredients. Mix the kimchi ingredients with the salted and dried Chinese cabbages.

In Jeonbuk, thin chal tteok (glutinous rice cake), borigae tteok (barely cake), and tteok jaban are well renowned. The last one in particular is made by the mixing of pepper paste with glutinous rice powder to make dough, frying it in a round and thin shape, then dipping this in the boiled seasoning paste that is made from soy sauce, sugar and fine pepper powder. It is usually consumed as a side dish.

Jeonju bibimbap
(Jeonju-style Bibimbap)*

Ingredients

Rice 540g, sliced raw beef (or stir-fried beef)150g, beef broth 800ml, bean sprouts 100g, water parsleys 100g, young pumpkins 200g, bellflowers 100g, fiddleheads 150g, dried shiitake mushrooms 10g, radishes 80g, cucumbers 70g, carrots 70g, yellow mung-bean jelly 150g, eggs 400g, chapssal gochujang (red hot pepper paste made with glutinous rice) 70g (or 4 tablespoons), fried kelps, pine nuts, cooking oil

For the seasoning for the sliced raw beef soy sauce 1 teaspoon, refined rice wine 1 teaspoon, sesame oil 1 teaspoon, crushed garlic, sesame seeds, sugar,

For the seasoning for vegetables (bean sprouts, water parsleys, young pumpkins, and bellflowers) salt, crushed garlic, sesame salt, sesame oil

For the seasoning for the fiddleheads and shiitake mushrooms

soy sauce, crushed garlic, sesame salt, sesame oil

For the seasoning for the radishes red hot pepper powder, salt, crushed garlic, crushed ginger

Cooking Directions

1 Cook the rice in the beef broth. Spread the cooked rice on a wide plate to allow to cool.

2 Season the sliced raw beef with the beef's seasoning.

3 Parboil the bean sprouts and water parsleys in boiling water and season with the seasonings for vegetables.

4 Slice the young pumpkins into thin pieces, salt them and squeeze the excess water out.
Cut the bellflowers into thin pieces, knead with salt to eliminate the bitter taste and soften the texture. Wring out the excess water. Mix with the seasoning and stir-fry them in cooking oil.

5 Soften the fiddleheads in water for one to two hours and parboil until the stems become softened. Cut them into short pieces. Soften the dried shiitake mushrooms in water, slice into thin pieces and stir-fry along with the seasoning.

6 Slice the radishes into thin pieces, season with the seasoning. Finely slice the cucumbers and carrots.

7 Slice the yellow mung-bean jelly into thin layers.
Fry the egg white and yolk separately into very thin layers. Cut up the fried kelp into short pieces.

8 Put all the prepared ingredients on top of the rice in a plate in a circular way.
Top with red hot pepper paste.

9 You can also add a raw egg and sprinkle pine nuts according to taste.

Note

Jeonju has long produced good-quality bean sprouts since it has clear water and fine weather. The key factor to determine the taste of Jeonju bibimbap is the sliced raw beef. People are used to eating bibimbap along with soybean sprout soup, stir-fried red hot pepper paste, sesame oil and nabak kimchi (or radish water kimchi).

---

* Bibimbap is known for being Korea's representative healthy food.

Hwangdeung bibimbap
(Bibimbap with Sliced Raw Beef and Vegetables)

Ingredients

Rice 360g, sliced raw beef 200g, bean sprouts 100g, spinaches 80g, yellow mung-bean jelly 80g, water for cooking the rice 470ml, eggs 200g, dried laver powder, a pinch of salt
For the seasoning for the sliced raw beef
soy sauce 2 tablespoons, crushed garlic 1 tablespoon, sesame oil 1 tablespoon, sugar 1 tablespoon, red hot pepper powder 2 teaspoons
For the seasoning sauce soy sauce 4 tablespoons, chopped green onions 2 tablespoons, crushed garlic 1 tablespoon, sesame oil 1 tablespoon, red hot pepper powder 2 teaspoons

Cooking Directions

1 Put the rice and water in a pot and cook.
2 Slice the beef into pieces of 5×0.3×0.3cm and mix well with the seasonings.
3 Parboil the bean sprouts and spinaches separately. Season the spinaches with salt.
4 Slice the yellow mung-bean jelly into pieces of 5×0.3×0.3cm.
5 Make the seasoning sauce.
6 Mix the parboiled bean sprouts and the seasoning sauce with the rice.
   Place them on a plate and add the spinach salad and sliced raw beef on top.
7 Garnish with pieces of dried laver, fried egg white and yolk strips, and yellow mung-bean jelly.
   Add sesame oil according to taste.

Aejeo jjim
(Steamed Baby Pork)

Ingredients

Baby pork 4kg, dried tangerine peel 500g, garlic 300g, ginger 200g, ginseng 5 pieces, gingko nuts 100g, chestnuts 100g, jujubes 80g, green onions 70g, onions 160g, refined rice wine 100g, water, red hot pepper paste mixed with vinegar.

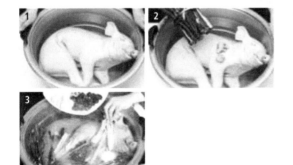

Cooking Directions

1 Soak the baby pork in cold water for 1-2 hours to remove the blood.
2 Pour enough water over the pork to cover it and add the dried tangerine peel, ginseng, garlic, ginger and refined rice wine. Bring them to the boil for 2 hours.
3 When everything is completely cooked, add chestnuts, gingko nuts, jujubes, onions, and green onions and keep boiling for a while.

Note

Back in the era when meat was considered a delicacy, a baby pig that was born defective was used to make the steamed dish above. Only the baby pork that is less than one month old and has consumed only its mother's milk was used for this dish. They cut the cleanly prepared baby pork into four to six pieces to steam for about two hours.

Hongeo jjim
(Steamed Skate)*

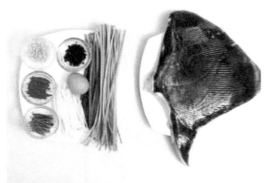

Ingredients

Skate 1kg, bean sprouts 80g, water parsleys 60g, leeks 30g, eggs 50g, red hot peppers 15g,
young green hot peppers 15g, manna lichen mushrooms, a pinch of pine nuts, salt, a little refined rice wine
For the seasoning sauce soy sauce 3 tablespoons, red hot pepper powder 2 teaspoons,
chopped small green onions 2 teaspoons, crushed garlic 1 teaspoon, sesame seeds, sesame oil

Cooking Directions

1 Make some slits on the skates and sprinkle salt and refined rice wine on them.

2 Put in a heated steamer and steam the skate.

3 Cut off the heads and tails of the bean sprouts. Cut the leeks and water parsleys into 5cm long pieces.

4 Parboil the bean sprouts, leeks and water parsleys in salty boiling water.

5 Fry the egg white and yolk separately and slice into 3cm long pieces.

6 Slice the red hot peppers and young green hot peppers into 3cm long pieces.
Soften the manna lichen mushrooms in water and also slice into pieces.

7 Put the parboiled vegetables, egg white and yolk, sliced peppers
and manna lichen mushrooms onto the steamed skate. Serve it with the seasoning sauce.

Note

You can also splash the seasoning sauce over the skate and
steam it along with the vegetables.

* The people in Jeollado enjoy grilled or fried skate along with fried potato chips. It is better to use a
light seasoning mixture.

Daehap jjim
(Steamed Clams)

Ingredients

Fresh clams 2kg, tofu 170g, beef 50g, eggs 200g, manna lichen mushrooms 5g, red hot peppers 60g, young green hot peppers 60g, flour 2 tablespoons
For the seasoning soy sauce 1/2 tablespoon, chopped green onions, crushed garlic, sugar, a pinch of sesame salt, sesame oil

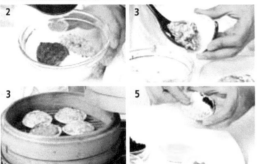

Cooking Directions

1. Soak the clams in salty water to remove any sediment. Wash them thoroughly.
2. Separate the flesh from the shells. Chop up the flesh, tofu and beef and mix well with the seasoning.
3. Clean the clam shells and remove the excess moisture. Stuff the ingredients from point 2 into the shells and sprinkle with flours. Brush one egg yolk over them and then steam.
4. Boil the rest of the eggs. Separate the white and yolk and put through the sieve individually to make powder. Finely chop up the manna lichen mushrooms, red hot peppers, and young green hot peppers.
5. Sprinkle the chopped ingredients from point 4 over the steamed clams from point 3.

Note

Steamed clam is best suited for spring and autumn. This dish is cooked by mixing beef and mushrooms along with clam flesh altogether and stuffing the mixture into the clam shell. In the 『Jeungbo sallim gyeongje (A Korean book on agriculture during the Joseon Dynasty)』, this is called 'Daehapjeung' while 『Our Food』, 『Home Cooking Around the World』, 『Korean Food』, and 『Traditional Korean Food』 call it 'Daehap jjim'. Moreover, 『Sallim gyeongje』 introduces the method for cooking clams, saying that "it is advised to place the dried clam flesh on top of rice, instead of eating them as they are or making a clam soup. They can also be salted and fermented as a sauce." It is said that the steamed clam was always presented during parties.

Gam danja
(Persimmon Rice Balls)

Ingredients

Persimmons 1.3kg, glutinous rice powder 500g, ginger 50g, powdered red beans 3 cups, sugar 5 tablespoons, water, a pinch of salt

Cooking Directions

1 Remove the stalks of the persimmons. Wash well. Boil thoroughly and sieve them to shake off the excess water. Crush them and place the crushed persimmon in another bowl.
2 Slice the ginger into thin layers. Boil them to make ginger broth.
3 Mix the crushed persimmons, sugar and salt well and pour the ginger broth over them in a pot.
4 Add the glutinous rice powder into the broth from point 3 and bring them to the boil, stirring fast.
5 When the glutinous rice powder is cooked and the whole ingredients become batter, remove the pot from the heat and then allow to cool.
6 Cut out the batter into dough like balls of an edible size. Coat them with the powdered red beans.

Note

You may choose to mix the glutinous rice powder, persimmon juice and ginger juice and steam them and then cover them with the powdered bean. Gam danja is one of the traditional Korean rice cakes made of glutinous rice and persimmons. While other ordinary rice cakes stale easily, gam danja keeps its taste for a long time. The soft texture of this healthy food also helps with digestion.

Seop jeon
(Pan-fried Chrysanthemum)

Ingredients

Glutinous rice powder 300g, water 100ml, soju (distilled liquor) 2 tablespoons, sugar 75g, cooking oil
For the garnish chestnuts 30g, jujubes 20g, manna lichen mushrooms 10g,
yellow leaves of chrysanthemum 30
For the syrup sugar 75, water 100ml

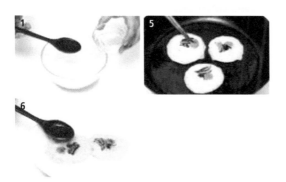

Cooking Directions

1 Combine the soju, water, and glutinous rice powder and mix them well.
2 Peel the chestnuts and slice into thin pieces. Peel and deseed the jujubes.
   Slice them into thin pieces. Soften the manna lichen mushrooms in water. Slice into thin pieces.
3 Coat the chrysanthemum leaves with the glutinous rice powder.
4 For the syrup, put the sugar in a pan, pour in the water and boil it until the quantity is reduced by half.
5 Line a heated pan with cooking oil and scoop the batter from point 1 and drop it into the pan.
   Add the sliced chestnuts, jujubes, manna lichen mushrooms, and chrysanthemum leaves on top of
   the batter. Pan-fry them.
6 Coat them with syrup when they are still hot.

Jeonju gyeongdan
(Jeonju-style Sweet Rice Balls)

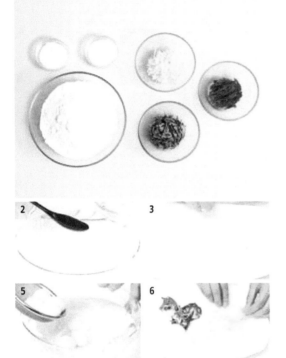

Ingredients

Glutinous rice 900g, chestnut strips 1/2 cup, jujube strips 1/2 cup, dried persimmon strips 1/2 cup, sugar 75g, water 150ml, salt 1 tablespoon

Cooking Directions

1 Wash the glutinous rice well and soften in water for more than 5 hours.
Add the salt and the grind and sieve the rice.
2 Pour boiling water into the glutinous rice powder little by little and mix well.
Beat the batter until it becomes soft.
3 Wrap the batter with a clean wet cotton cloth. Make balls out of the batter into the size of a chestnut.
4 Put the sugar and water in a pot to boil. Add the balls from point 3 and keep boiling them.
5 When the balls begin to float to the surface, take them out and wash in cold water.
Sieve them to drain the water.
6 Spread the strips of chestnuts, jujubes and dried persimmons separately on a plate.
Place the balls on them and cover with the extra strips.

[Jeonnam]

Jeonnam's food is very diverse. People who live near the Southwestern seashore usually eat fishery products while those who live in the Northwestern mountainous area frequently consume wild herbs and vegetables. Skate ray has been considered as a delicacy in Jeonnam and it is always served at the major events such as weddings.

Jeonnam kimchi is made mostly with: Chinese cabbage; radishes; young radishes; cucumbers; mustard leaves and stems; Korean lettuce; green onions; unripe peppers; unripe garlic and green algae. Salted seafood and pepper powder are also often used while the broth is small in its volume for this dish.

Jeonnam's tteok includes a considerable amount of salt and sugar along with the tteok powder. The blue color is produced by including ramie fabric leaf or mugwort.

Daetong bap
(Rice in a Bamboo Stalk)

Ingredients

Non-glutinous rice 150g, brown rice 30g, barley rice 30g, black rice 10g, chestnuts 130g, gingko nuts 30g, jujubes 16g, some water

Cooking Directions

1 Combine the non-glutinous rice, glutinous rice, brown rice, barley rice, and black rice, mix and wash well. Soften them in water overnight.
2 Drain and stuff the softened rice from point 1 in a 60 percent part of a bamboo stalk. Pour in the water until it reaches 1cm above the rice.
3 Add the chestnuts, gingko nuts, jujubes, and pine nuts to top of the rice. Cover with hanji (a traditional Korean paper).
4 Place the bamboo stalk in a pot. Fill with water until it reaches half the height of the bamboo stalk. Steam for 40 minutes.
5 Leave the rice settle in its own heat from five to ten minutes.

Yukhoe bibimbap
(Bibimbap with Sliced Raw Beef)

Ingredients

Boiled rice 840g, beef 200g, bean sprouts 100g, spinaches 100g, young pumpkins 100g,
pine mushrooms 100g, sliced radishes 100g, lettuce 5g, eggs 200g, dried laver powder 5g,
red hot pepper paste 4 tablespoons, red hot pepper powder 1 tablespoon, chopped green onions 6
tablespoons, crushed garlic 3 tablespoons, salt 1 tablespoon, light soy sauce , sesame oil, sesame salt

Cooking Directions

1. Slice the beef along the opposite direction of the texture into pieces (5x0.2x0.2cm).
   Season with sesame oil and sesame salt.
2. Clean the bean sprouts. Add salt into a pot and parboil the bean sprouts.
   Season well with the chopped green onions, crushed garlic, salt, and sesame oil.
3. Parboil the spinaches and rinse in cold water. Wring out the excess moisture.
   Season with the chopped green onions, crushed garlic, light soy sauce, and sesame oil well.
4. Parboil the softened fiddleheads.
   Season with light soy sauce, sesame oil, crushed garlic, and sesame salt and stir-fry them.
5. Slice the young pumpkins into pieces (5x0.2x0.2cm). Shred the pine mushrooms by hands
   and stir-fry with salt and sesame oil. Slice the lettuce into 0.2cm wide pieces.
6. Add the red hot pepper powder, crushed garlic, salt, sesame oil, and sesame salt
   to the sliced radishes and mix well.
7. Place the boiled rice in a bowl. Arrange the prepared ingredients on top and top with the egg yolk,
   sesame salt, red hot pepper paste, and dried laver powder on the boiled rice.

Naju gomtang
(Naju-style Bone Soup)

Ingredients

Ox bones, beef (shank, brisket) 150g, radishes 200g, onions 50g, green onions 35g,
garlic 15g, eggs 50g, crushed garlic, red hot pepper powder, salt, sesame oil, sesame seeds, water

Cooking Directions

1 Put the ox bones in a pot and pour enough water over them to cover.
Bring them to the boil for a long time. Pour the broth into another bowl.
2 Pour water over the ox bones and bring them to the boil again,
this time until the color becomes a clear white. Mix the broth from point 1 and 2.
3 Add the beef, radishes, onions, some of the green onions and garlic into the combined broth
and bring to boil.
4 When the meat is cooked, take out and slice into thin pieces. Chop up the remaining green onions.
5 Sieve the broth from point 3 to make it clear.
6 Fry the egg white and yolk individually into very thin layers and slice them into pieces of (5x0.2x0.2cm)
7 Place the broth from point 5 in a bowl. Place the sliced meat into the bowl. Garnish with the
chopped green onions, crushed garlic, fried egg white and yolk strips, sesame seeds, sesame oil, and
red hot pepper powder. Serve with salt.

Note

Naju-style gomtang tastes especially great. Unlike seolleongtang
(ox bone soup) or other soup, it doesn't require other innards
of the beef. When briskets, beef shanks, or chunks are added
and boiled together, the broth becomes much clearer and
tastier.

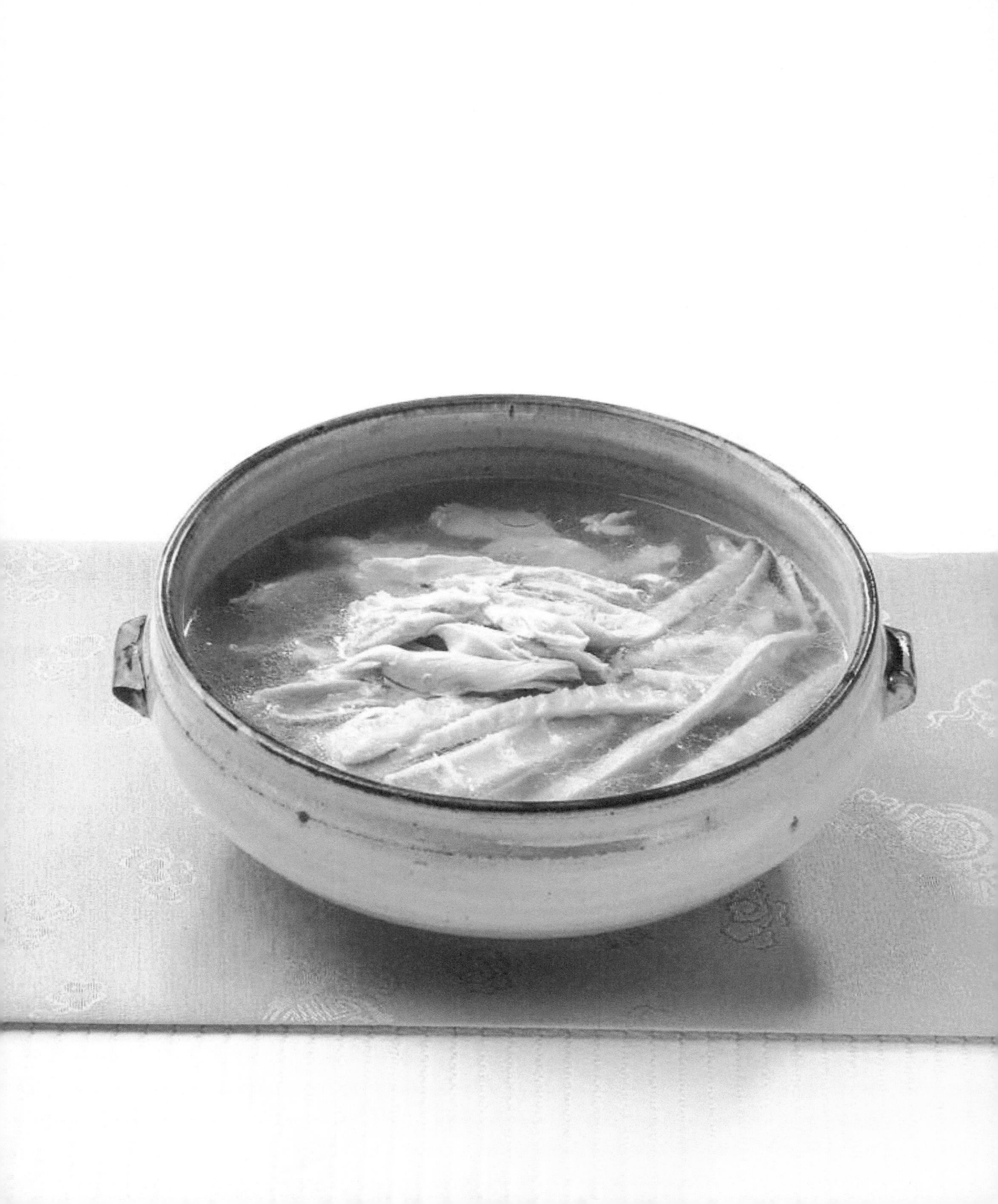

Juksun tang
(Bamboo Shoots Soup)

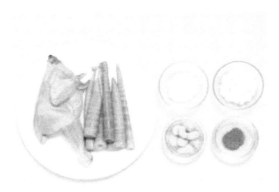

Ingredients

Bamboo shoots 400g, a young chicken (weighing approx 800g), glutinous rice 2 tablespoons, garlic 20g, rice water 600ml, water 2.4L, salt 1 teaspoon, a pinch of black pepper

Cooking Directions

1. Cutoff the tail of a young chicken and remove the giblets.
   Wash thoroughly inside and out to remove all the blood and then drain.
2. Boil the bamboo shoot in rice water. Soak in tepid water to remove the bitter taste.
3. Wash the glutinous rice and then soften in water.
4. Stuff the young chicken with the softened glutinous rice and garlic.
   Sew the body cavity with cotton threads.
5. Put the prepared chicken in a pot with the bamboo shoots, pour enough water in to cover and bring to the boil.
6. Once the chicken is completely cooked, take out the chicken and the bamboo shoots.
   Season the broth with salt and black pepper.
7. Shred the chicken meat and bamboo shoots. Put them in a bowl and pour over the broth.

Note

Naju-style gomtang tastes especially great. Unlike seolleongtang (ox bone soup) or other soup, it doesn't require other innards of the beef. When briskets, beef shanks, or chunks are added and boiled together, the broth becomes much clearer and tastier.

Mareun hongeo jjim
(Steamed Dried Skate)

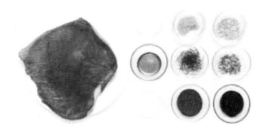

Ingredients

Skate 1kg

For the seasoning sauce soy sauce 3 tablespoons, red hot pepper powder 2 tablespoons, chopped small green onions 1 teaspoon, crushed garlic 1 tablespoon, sugar 1 teaspoon, sesame oil 1 teaspoon, sesame salt 1 teaspoon, salt, sliced red hot peppers

Cooking Directions

1 Place the skate in a plate and dry in the shade. Cut it into three pieces and then steam them.
2 Make the seasoning sauce with the ingredients above.
3 When the skate is completely cooked, place in a plate and cover with the seasoning sauce.

Kkomak muchim
(Seasoned Small Clams)*

Ingredients

Small clams 400g, water, salt
For the seasoning soy sauce 2 tablespoons, red hot pepper powder 1 tablespoon,
chopped green onions 2 tablespoons, crushed garlic 1 tablespoon, crushed ginger 1/2 tablespoon,
sugar 1 teaspoon, sesame oil, sesame seeds, sliced red hot peppers

Cooking Directions

1 Rub the small clams in running water and then soak in salt water for 2 hours
to remove any sediment.
2 Mix the ingredients to make the seasoning sauce.
3 Put the small clams into boiling water. Reduce the heat and keep on the boil while stirring.
Take out the clams from the pot before the shells are fully opened.
4 Remove one side of the shells and then place them on a plate.
5 Splash the seasoning sauce on the clams.

---

* This dish is very popular among the Korean people.

Bajirak hoemuchim
(Seasoned Manila Clams)

Ingredients

Short-necked clam flesh 300g (or 1 1/2 cups), young pumpkins 400g, cucumbers 145g, water parsleys 80g, carrots 50g, small green onions 30g
For the red hot pepper paste sauce with vinegar red hot pepper paste 3 tablespoons, vinegar 3 tablespoons, red hot pepper powder 2 tablespoons, sugar 2 tablespoons, crushed garlic 1 tablespoon, sesame seeds 1 tablespoon, salt 1 teaspoon,

Cooking Directions

1 Parboil the clams in boiling water.
2 Slice the young pumpkins and carrots into pieces (5x0.3x0.3cm).
   Peel and deseeds the cucumbers and cut diagonally into 0.3cm long pieces.
3 Cut the water parsleys into 5cm long pieces and parboil in boiling water.
4 Cut the small green onions into 2cm long pieces. Cut the white parts into two pieces.
5 After making the red hot pepper paste sauce with vinegar, add the young pumpkins, cucumbers, carrots, water parsley, small green onions and clam flesh and mix well.

Note

As an alternative you can choose other clams

Gim bugak
(Fried Laver)

Ingredients

Dried laver 200g (or about 100 sheets of dried laver), glutinous rice powder 500g, broth (ingredients: anchovy, kelp, shiitake mushrooms, water 1.6L), sesame seeds 90g, cooking oil 3 cups, salt, and soy sauce

Cooking Directions

1 Prepare the dried laver.
2 Add the glutinous rice powder to the broth made from anchovy, kelp, shiitake mushrooms and season with salt and soy sauce. Stir the batter with a wooden scoop to make sticky glutinous rice glue.
3 Place a sheet of dried laver on a kitchen board. Apply the glue from point 2. Sprinkle the sesame seeds on it. Cover with another sheet of dried laver. Dry in the sun.
4 When they are thoroughly dried, cut them into four pieces. Wrap them with a plastic bag to preserve. Fry them just before eating by slightly/fast frying them over a low heat.

Deulkkaesongi bugak
(Fried Perilla Clusters)

Ingredients

Perilla clusters 30, glutinous rice powder 100g, water 400ml, salt 2 teaspoons, cooking oil

Cooking Directions

1 Wash the perilla clusters well and sieve them to remove the moisture.

2 Add the glutinous rice powder to the water and season with salt.
  Stir the batter with a wooden scoop or spoon to make sticky glutinous rice glue.

3 Apply the glue to the perilla clusters. Arrange them separately in a tray.
  Dry in the sun. Repeat this process (from applying glue to dry in the sun) three times.

4 When they are completely dried, wrap them with a plastic bag to preserve.

5 Fry them just before eating by slightly/fast frying them over a low heat.

Acacia bugak
(Fried Acacia)*

Ingredients

Acacia flowers 300g, glutinous rice glue 1 cup, cooking oil

Cooking Directions

1 Wash the acacia flowers well and shake in a sieve to remove the moisture.

2 Apply the glutinous rice glue to both sides of the acacia flowers. Arrange them separately in a tray.
Completely dry in the shade. Applying the glue again and dry in the sun.

3 Slightly fry them over a low heat.

* It is very popular as a dessert since it tastes sweet and smells great.

Kkaetip bugak
(Fried Sesame Leaves)

Ingredients

Sesame leaves, flour, grain syrup (or dextrose syrup), chopped green onions, crushed garlic,
crushed ginger, salt, sesame oil, cooking oil

Cooking Directions

1 Wash the sesame leaves well. Soak in the salt water for about 10 minutes. Rinse with water.

2 Mix with flour and steam for 30 minutes.

3 Completely dry the steamed sesame leaves and then fry in cooking oil.

4 Mix the grain syrup (or dextrose syrup), chopped green onions, crushed garlic, crushed ginger, salt,
and sesame oil and bring them to the boil.

5 Soak the fried sesame leaves in the syrup and allow to cool.

Ujjiji
(Decorated Rice Cake)

Ingredients

Glutinous rice powder 500g, water for batter 130ml, powdered red beans 100g, jujubes 25g,
manna lichen mushrooms 10g, sugar 2 tablespoons, honey 1 tablespoon, salt 1 teaspoon, cooking oil

Cooking Directions

1. Mix the glutinous rice powder with hot water.
2. Put the red beans and water in a pot, bring to the boil, and sieve them to remove the excess water.
   Add sugar and honey and mix well. Make dough and then roll to make some small balls.
3. Peel and deseed the jujubes. Cut them into 0.1cm thick pieces.
   Rub the manna lichen mushrooms with salt and wash them. Roll them up and slice in thin strips.
4. Take the balls made out of the glutinous rice batter. Line a pan with cooking oil.
   Place the balls in the pan in a circle. Fry them over a low heat and squash them down with a spoon.
   Spoon out a lump of the powdered red beans and place this into the middle of the ball.
   Roll up with the filling inside. Slightly press both ends.
5. Garnish the balls with jujubes and manna lichen mushrooms in a flower form.

Moyakgwa
(Fried Honey Cookies)

Ingredients

Flour 1kg, ginger 20g, refined rice wine 1 cup, cooking oil 1/2 cup, sesame oil 1/2 cup, cinnamon powder 2 tablespoons, salt 1 tablespoon, some pine nuts,
For the syrup dextrose syrup 2 cups, sugar 2 tablespoons, water 200ml

Cooking Directions

1 Mix the flour, cinnamon powder, salt, sesame oil, and oil well and sieve them while rubbing by hand.

2 Grind the ginger to make juice and then mix with the refined rice wine.

3 Mix 1 and 2 to make rough dough.

4 Roll the dough out into a 0.5 cm layer and cut into 3x3 cm.
  Score at the edge or punch the middle of each piece.

5 Fry the dough for 10 minutes at 150℃, 15 minutes in 100℃ and then 5 minutes in 150℃.

6 Mix the dextrose syrup, sugar and water well and bring them to the boil to make syrup.
  Apply the syrup to the fried cookies made from point 5.

Saenggang jeonggwa
(Candied Ginger)

Ingredients

Ginger 100g, dextrose syrup 2 cups, sugar 3 tablespoons, salt 1/2 teaspoon

Cooking Directions

1 Peel the ginger and slice it into thin layers
2 Put some salt and the ginger into boiling water to parboil.
  Wash the parboiled ginger and rinse in cold water. Place in a tray.
3 Mix the dextrose syrup, sugar, and water in a pot and bring them to the boil over a high heat.
  Add the ginger and braise over a low heat with the lid kept open. Skim off the foam while it is boiling.
4 When the food is completely braised. Take out one by one and allow to cool.

Note

Jeonggwa is also called jeongwa, which is a type of sticky candy that tastes sweet. It is made by braising vegetable roots, fruits, stalks, or berries such as lotus roots, bellflowers, ginger, ginseng, yellow quinces, citrons, apples or the fruit of Chinese matrimony vines and then adding honey or sugar. You can dry the braised ginger to make it into a dry candy.

[Gyeongbuk]

Living up to its reputation as a conservative area, Gyeongbuk food is very traditional, and this has developed into 'local' food as we know it today. The Andong cultural block is characterized by the religious ceremonial food of the Confucian denomination, while the Gyeongju region offers palatial foods and religious ceremonial food, affected by the Buddhist culture of Shilla

Including the rice produced from near the Nakdong River and from the wide and rich inland plain, the supply of diverse vegetables and meat is effective during each season. Moreover, as the longest seashore in Korea and the East Sea are also present here, fishery products and stored foods are well developed in their production. In the mountainous areas, potato, sweet potato, buckwheat and dotori muk (acorn starch jelly) are also well developed. Food here tends to be very spicy, salty and devoid of ornaments.

Daege bibimbap
(Bimbap with Steamed Snow Crab)

Ingredients

Boiled rice 840g, snow crabs 2, cucumbers 150g, young pumpkins 120g, carrots 120g,
bellflowers 80g, eggs 50g, dried laver 2g, salt 1 tablespoon, cooking oil 1 teaspoon,
sesame oil 1/2 tablespoon, crushed garlic 1 tablespoon,sesame salt 1 tablespoon, a pinch of sugar

Cooking Directions

1 Wash the snow crabs thoroughly and lay them on their backs in a pot. Steam them for 10 minutes.

2 Peel the young pumpkins and cucumbers and slice them into 5cm long pieces.
Salt them to remove the excess moisture and stir-fry in a pan.

3 Cut the carrots into strips (5×0.2×0.2cm). Put some salt and the carrots in boiling water
to parboil them. Take out the carrots from the pan and season with salt and sesame oil.
Stir-fry in a pan.

4 First cut the bellflowers into 5cm long pieces and then cut into thin pieces.
Rub with salt to eliminate the bitter taste. Parboil them in boiling water. Take out the bellflowers
from the pan and season with the sugar, crushed garlic, sesame salt, and sesame oil and stir-fry in a pan.

5 Fry the egg white and yolk separately into thin layers and slice them into thin pieces (5×0.2×0.2cm).

6 Grill the dried lavers, and slightly break them.

7 Remove the lids and innards of the steamed snow crabs. Tear off the flesh from the legs.

8 Scoop the boiled rice in a bowl. Place the vegetables on top from point 2, 3, and 4
as well as the crab flesh. Garnish with the dried laver and the egg white and yolk strips.

Jobap
(Millet rice)

Ingredients

Rice 270g, glutinous millet 75g, water 470ml

Cooking Directions

1 Wash the rice well and soften in water for 30 minutes.
2 Wash the glutinous millets and soften in water for 30 minutes. Remove the water.
3 Pour the water over the softened rice in a pot and bring it to the boil.
   When it reaches the boil, add the glutinous millet and bring them to the boil again.
4 Let the rice settle in its own steam.

Note

You can mix the rice and millets from scratch and bring them to the boil together. You can also add red beans, soybeans or Indian millets.

Daegu yukgaejang
(Daegu-style Spicy Beef Soup)*

Ingredients

Beef (brisket) 600g, radishes 200g, mung-bean sprouts 300g, taro stalks 200g, green onions 70g, water 3L, red hot pepper powder 2 teaspoons, sesame oil 1 teaspoon, salt 1 teaspoon
For the seasoning light soy sauce 2 tablespoons, chopped green onions 1 tablespoon, crushed garlic 1 tablespoon, sesame salt 1 teaspoon, a pinch of black pepper

Cooking Directions

1 Cut the radishes and beef into bite sizes. Bring the radishes and beef to the boil over a low heat.
2 Nip off the tails of the mung-bean sprouts.
  Parboil them in boiling water and then wash in cold water. Wring out the moisture.
3 Parboil the taro stalks in boiling water and wash in cold water.
  Wring out the moisture and cut into 10cm long pieces.
4 When the beef and radishes from point 1 are cooked, take out them from the broth.
  Slice the beef into thin layers. Slice the radishes into rectangular pieces of (2×2×0.5cm).
  Season with the seasoning.
5 Put the parboiled taro stalks and green onions in the broth and bring them to the boil.
  When they begin to boil, add the mung-bean sprouts and seasoned beef and radishes
  and bring them to the boil once again.
6 Combine the sesame oil, red hot pepper powder and 2 spoons of the broth from point 11
  and mix well. Put this back into the broth and stir well. Season with salt.

Note

The spicy beef soup has been known nationwide by refugees since the Korean War in 1950.

* Many Koreans enjoy this kind of soup dish

Dubu saengchae
(Tofu Salad)

Ingredients

Radishes 500g, tofu 120g, salt 1 teaspoon
For the seasoning: red hot pepper powder 2 teaspoons, salt 1 teaspoon,
sesame oil 1 tablespoon, sesame salt 1 tablespoon

Cooking Directions

1 Finely slice the radishes into strips (5×0.2×0.2cm) and salt them. Wring out the water.
2 Mesh the tofu with the back of a knife.
  Wrap it with a cotton cloth and wring out it to remove the excess moisture.
3 Mix the sliced radishes and tofu with the red hot pepper powder well.
  Season them with the salt, sesame salt, and sesame oil.

Gajami jorim
(Braised Sole with Vegetables)

Ingredients

Dried soles 200g, a pinch of sesame seeds

For the seasoning sauce Dried hot peppers 2, soy sauce 4 tablespoons,
red hot pepper paste 1 tablespoon, red hot pepper powder 2 tablespoons, dextrose syrup 1/2 cup,
sugar 1 tablespoon, water 200ml, crushed garlic 1 teaspoon, some cooking oil

Cooking Directions

1 Cut the dried hot peppers into 1cm long pieces
2 Make the seasoning sauce with the ingredients. Bring to the boil and allow to cool.
3 Cut the dried soles into bite sizes.
  Fry in cooking oil until they are crispy. Allow them to cool.
4 Season the fried soles with the seasoning sauce, and sprinkle the sesame seeds on top of them.

Jaban godeungeo jjim
(Steamed Mackerel)

Ingredients

Salted mackerel 400g, young green hot peppers 30g, green onions 35g,
a pinch of black sesame seeds and chopped red hot peppers, rice water 1L

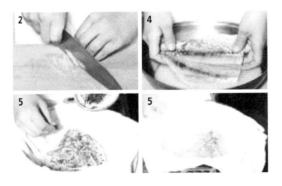

Cooking Directions

1 Cut the young green hot peppers into two pieces.
Deseed and slice them into thin pieces (3×0.1×0.1cm).
2 Choose only the white parts of the green onions and slice them into pieces (3×0.1×0.1cm).
3 Remove the tails and bones from the salted mackerel.
Soak them in the rice water to remove the salty taste.
4 Place a cotton cloth on a bottom of a steamer. Place in the mackerel and garnish
with the young green hot peppers, green onions, sliced red hot peppers and black sesame seeds.
Steam for 10 minutes.

Note

The jaban godeungeo (gan godeungeo, or salted mackerel)
originated from the mackerel enjoyed in Andong province.
They used to salt the mackerel caught in the East Sea in order
to prevent them spoiling during transportation to shore. The
salted mackerel became a specialty of this region because it
was especially delicious. The steamed mackerels are usually
served with lettuce, butterbur leaves, kelp and soybean paste
sauce.

Kong tang
(Soft Tofu, Soybean Stew)

Ingredients

Fresh soybean powder 120g, small green onions 20g, leeks 20g, water 600ml

For the salty water water 2 tablespoons, salt 1 tablespoon

Cooking Directions

1 Cut the leeks and small green onions into 3cm pieces.

2 Mix the soybean powder and some water (200ml) well.

3 Boil the rest of the water (400 ml) and add the mixture from point 2 and bring them to the boil again stirring well until the stink from the soy bean is removed.

4 Add the salty water into point 3.

5 When the soy beans are softened and clotted like raw tofu, add the leeks and small green onions.

Gam gyeongdan
(Persimmon Rice Balls)

Ingredients

Glutinous rice powder 550g, mung-bean powder 115g, soft persimmons 2, ginger 50g, salt 1 1/2 teaspoon, a pinch of cinnamon powder, water 300ml

Cooking Directions

1. Slice the ginger into thin layers. Put them in a pot and pour 1 cup of water over them. Bring them to the boil to make ginger juice.
2. Make several slits on the soft persimmons and peel them. Pour water 100ml over the soft persimmons and bring them to the boil. Put them in a sieve to shake off the water.
3. Combine the glutinous rice powder, salt 1 teaspoon, ginger juice 2 tablespoons, soft persimmon juice 5 tablespoons from point 2, and cinnamon powder and mix well. Mix them with hot water to make the dough.
4. Make small dumplings (about the size of a chestnut).
5. Pour water in a pot and add 1/2 teaspoon of salt and bring it to the boil. When it begins to boil, put the dumpling from point 4 in one by one. When they begin to float, take out them and wash in cold water. Place them on a tray to remove the excess moisture.
6. Coat the dumplings with the mung-bean powder.

Hongsi tteok
(Rice Cake with Soft Persimmons)

Ingredients

Non-glutinous rice powder 1kg, soft persimmons 3, carrots 75g, sugar 150g, dextrose syrup 75g, salt 1 tablespoon, water 100ml

Cooking Directions

1 Remove the stalks of the soft persimmons. Make several slits on the soft persimmons and peel them. Pour water over the soft persimmons and bring them to the boil.

Place in a sieve to shake off the excess water.

2 Carve the carrots into beautiful shapes (flower shape etc.) and soften them in the dextrose syrup for an hour.

3 Combine the non-glutinous rice powder, soft persimmons and salt and mix well.

Sieve them and season with sugar.

4 Place a cotton cloth in a steamer and place in the mixture from point 3.

5 When the steam rises, cover with another cloth and steam for 15 minutes further.

6 Cut into bite sizes and place on a plate. Decorate with the carrots.

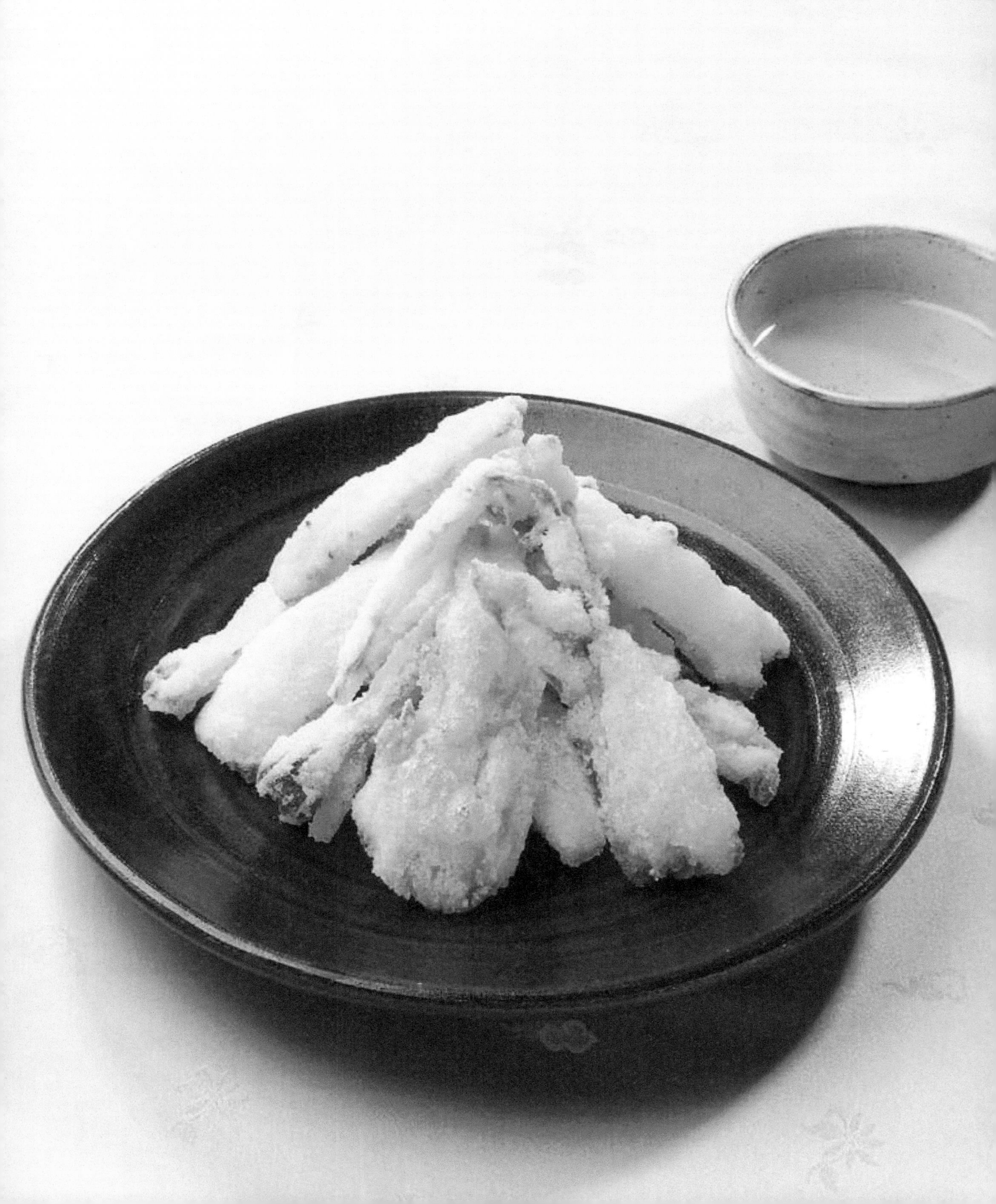

Seopsansam
(Fried Deodeok)

Ingredients

Deodeok roots (roots of the Codonopsis lanceolata) 200g, glutinous rice powder 50g, water 200ml, honey 2 tablespoons, salt 1 teaspoon, cooking oil

Cooking Directions

1 Peel the deodeok roots. Beat with a bat and soak in salty water. Remove the excess moisture.
2 Coat the deodeok with the glutinous rice powder.
3 Pour the cooking oil in a pot and fry the deodeok from point 2 at 160℃.
4 Serve with honey.

Note

Sprinkling sugar on the fried deodeok will also be good.

[Gyeongnam]

Gyeongnam is characterized by its balanced nutritional diet that includes fresh agricultural and fishery products. Sashimi, fried dishes, steamed dishes, stir fried foods and soups using fish are developed in a diverse manner, and salted seafood types are also well developed.

Kalguksu (hand-rolled noodle soup) is regarded as the best delicacy from this area and the broth is made with anchovies or sea shells. The inland plain area produces raw mountain vegetables during spring, cucumber, pumpkins, eggplants, pepper and tomato during summer and dried vegetables during winter. As the weather is often warm, salt is often used in order to prevent the foods from deteriorating. The salted anchovies, which are mostly supplied from the South Sea, are used mostly with all types of foods, including kimchi.

During large scale events, salad or sanjeok (meat and vegetables kebob) is made with fishery products (such as mussels, spiny top shells, abalone, octopus, shark etc.).

Farmers living here usually eat boiled potatoes, sweet potatoes, pumpkins and others, memil muk (buckwheat jelly), Dotori muk (acorn starch jelly), wheat cakes and bran cakes. Food here is also devoid of ornaments and is made simple.

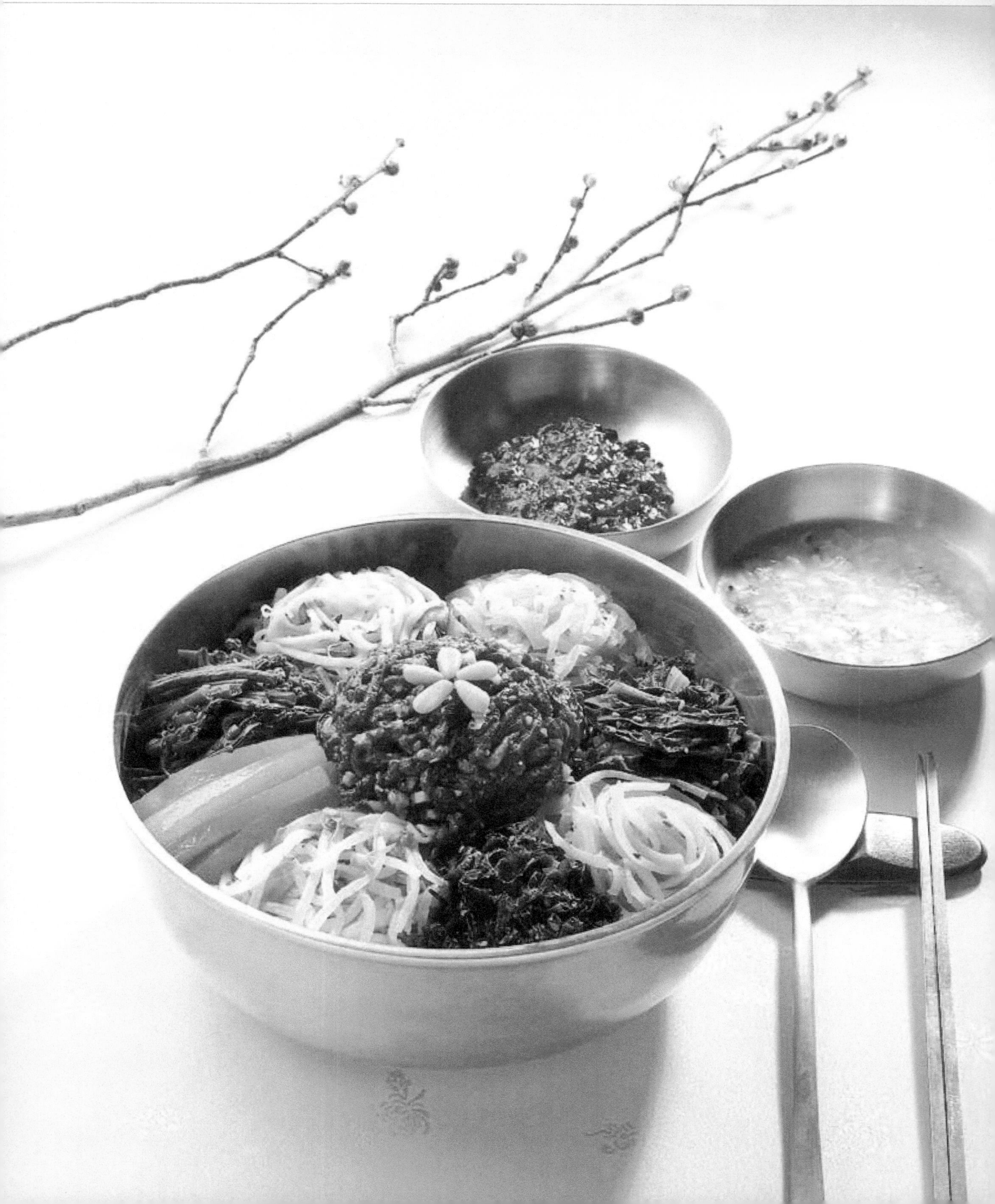

Jinju bibimbap
(Jinju-style Bibimbap)

Ingredients

Rice 360g, water 470 mL, mung-bean sprouts 130g, bean sprouts 130g, fiddleheads 100g, young pumpkins 100g, bellflowers 100g, beef 200g, yellow mung-bean jelly 100g, dried lavers 10g, radishes 100g, pine nuts 10g, red pepper paste with malt 2 tablespoon, light soy sauce 2 tablespoons, sesame salt 1/2 tablespoon, sesame oil 1/2 tablespoon

For the seasoning for the sliced raw beef sesame oil 2 tablespoons, sugar 1 tablespoon, crushed garlic 1/2 tablespoon, chopped green onions 1 tablespoon, sesame salt 2 teaspoons, a pinch of salt and black pepper

For the broth clams 130g, light soy sauce, water 100ml

Cooking Directions

1. Soften the rice in water for 30 minutes and bring it to the boil.
2. Slice the beef into thin pieces and season with the seasoning.
3. Nip off the heads and tails of the mung-bean sprouts and nip off the tails of the bean sprouts. Bring them to the boil.
4. Parboil the spinaches and fiddleheads separately in boiling water.
5. Slice the young pumpkins, radishes, and bellflowers into thin pieces (5×0.2×0.2cm), and parboil them in boiling water.
6. Tear the dried lavers with hands. Slice the yellow mung-bean jelly into thick pieces. (5×0.5×0.5cm)
7. Add the light soy sauce, sesame salt, and sesame oil into the ingredients from point 3, 4, 5, and 6 separately and mix well.
8. Wash the clams thoroughly. Put the clams and water in a pot and bring them to the boil. Season with light soy sauce to make the broth.
9. Place the boiled rice on a plate. Arrange the prepared ingredients (namul) from point 7 with the colors matching. Splash the broth from point 8 slightly. Place the seasoned raw beef on top of it.
10. Garnish the bibimbap with pine nuts. Serve with the broth and the malt red hot pepper paste.

Chungmu gimbap
(Chungmu-style Laver Rice Rolls)

Ingredients

Rice 360g, dried laver (seaweed) 8g, squids 200g, radishes 150g, water 470ml, a pinch of salt

For the seasoning of the squids red hot pepper powder 2 tablespoons, soy sauce 2 tablespoons, crushed garlic 1 teaspoon, chopped green onions 1 teaspoon, sesame salt 1/2 teaspoon, salt 1/2 teaspoon, sugar 1/2 teaspoon, sesame oil 1 teaspoon, a pinch of black pepper

For the seasoning of radishes salted shrimps 1 tablespoon, red hot pepper powder 2 1/2 tablespoons, crushed garlic 1 teaspoon, chopped green onions 1 teaspoon

Cooking Directions

1 Wash the rice thoroughly and soften it in water for 30 minutes.
2 Peel the squids and parboil in boiling water.
Cut into 2x4 cm long pieces and season with the ingredients.
3 Cut up the radishes diagonally. Salt slightly, wash and then remove the excess moisture.
Season with the seasoning.
4 Cut the dried laver into six pieces. Spoon out a lump of the boiled rice and place it into the middle of the dried laver. Roll up the dried laver with the rice inside. Serve with the prepared squids and radishes.

Note

In the past, women used to sell gimbap, squids and radish kimchi from a wooden bowl at the shuttle vessels, which were transported between Tongyeong and Busan; these were called 'Chungmu gimbap'. Chungmu gimbap (dried laver rice rolls) is also referred to as Halmae (grandmother) gimbap. It came about when rice and side dishes were consumed separately to prevent transmutation of the rice during summer time. Webfoot octopus was used originally, but now squids are used instead.

Majuk
(Yam porridge)

Ingredients

Rice 300g, yams 250g, water 1.6L, salt 1 teaspoon, some honey

Cooking Directions

1 Soften the rice well by soaking. Then drain and pour water over it and then finely grind.

2 Peel the yams and grate them.

3 Boil the ground rice thoroughly.

  Add the grated yam and keep boiling for a while. Season with salt.

4 Serve with honey.

Note

The porridge is also made from grated yam, mung-bean starch, and potato starch. You can also mix boiled yam and the softened rice to make the porridge.

Jinju naengmyeon
(Jinju-style Cold Buckwheat Noodles)

Ingredients

Buckwheat noodles (fresh noodles) 600g, radish kimchi 150g, beef (or pork) 150g, eggs 50g, pears 120g, a pinch of sliced red hot peppers, a pinch of pine nuts, some cooking oil, broth (made from seafood ingredients described below) 1.2L
For the seasoning for the beef soy sauce 1/2 tablespoon, chopped green onions 2 tablespoons, crushed garlic 1 teaspoon, some sesame oil, a pinch of sugar, a pinch of sesame salt, a pinch of black pepper
For the starchy sauce starch 1 teaspoon, water 1/2 tablespoon
For the seafood broth dried Pollack head, dried shrimps, dried mussels, some water

Cooking Directions

1 Put the ingredients for the seafood broth in a pot and bring them all to the boil.
2 Cut the beef into thin pieces and season with the seasoning. Soak in the beaten egg. Line a pan with cooking oil and pan-fry the beef. Cut it again into 1cm wide pieces.
3 Wring out the radish kimchi to remove the excess moisture. Peel the pears and cut into 0.5 thick pieces.
4 Combine the rest of the beaten egg and starchy sauce and mix well. Fry the mixture into very thin layers and slice into thin pieces (5x0.2x0.2cm).
5 Slice the red hot peppers into strips of 3-4cm.
6 Boil the buckwheat noodles and rinse them in cold water and wash several times. Coil the noodles into a bowl.
7 Place the prepared beef, radish kimchi, pears, fried egg strips, sliced red hot peppers, and pine nuts on the top of the coiled noodle. Pour over the seafood broth.

Note

As buckwheat was grown in the area around Mt. Jiri, the buckwheat noodle was enjoyed in this region. Where there is Pyeongyang naengmyeon in what is now North Korea, there is Jinju naegmyeon in the South.

Jaecheop guk
(Clear Shellfish Soup)

Ingredients

Shellfish 800g, leeks 20g, water 1.6L, salt 1 1/2 teaspoons

Cooking Directions

1 Soak the shellfish in salty water (using 1/2 teaspoon of salt) to remove any sediment.

2 Cut the leeks into 0.5 cm pieces.

3 Put the shellfish in a pot and pour water over them. When they begins to boil, take the flesh from the shells and season with salt (1 teaspoon). Add the leeks before turning off the heat.

Note

Shellfish in Korea can also be called gaeng shellfish, which mean the shellfish living in the river. You can also add red hot pepper powder or soybean paste.

Busan japchae
(Busan-style Stir-fried Noodles with Vegetables)

Ingredients

1/2 Octopus, mussels 110g, abalones 85g, clams 50g, onions 80g,
young green hot peppers 30g, glass noodles 50g, cooking oil 1 teaspoon
For the seasoning for the glass noodles soy sauce 1 tablespoon,
sugar 1/2 teaspoon, a pinch of sesame oil
For the seasoning for the japchae soy sauce 1 tablespoon, sesame salt 1 tablespoon, sesame oil 1
teaspoon, sugar 1 teaspoon, black pepper

Cooking Directions

1 Put the octopus in a pot and steam it. Cut up diagonally into pieces.
2 Trim the clams, abalone, and mussels and parboil them separately.
   Cut up diagonally into small pieces.
3 Slice the onions into 0.3cm thick pieces. Cut the young green hot peppers into two.
   Deseed them and slice again into the same size of pieces as that of the onions.
4 Soften the glass noodles in water. Parboil and cut up into pieces.
   Season with the seasoning for the glass noodles and stir-fry in a pan.
5 Line a pan with sesame oil and stir-fry the onions and young green hot peppers.
6 Combine the glass noodles, mussels, clams, abalone, octopus, onions, and young green hot peppers
   and season them with the seasoning for the japchae.

Eonyang bulgogi
(Eonyang-style Bulgogi)

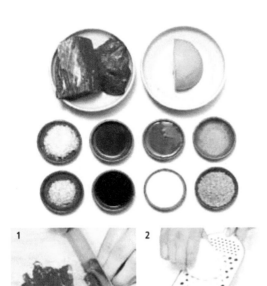

Ingredients

Beef 600g, pears 90g, a pinch of sesame seeds

For the seasoning sauce light soy sauce 1 1/2 tablespoons, sugar 1 1/2 tablespoons, chopped green onions 2 tablespoons, crushed garlic 1 tablespoon, honey 1 tablespoon, starch syrup 1 teaspoon, sesame oil 1 teaspoon, a pinch of ground black pepper

Cooking Directions

1 Slice the beef into thick pieces of 3x5 cm

2 Peel and deseed the pears and grate them to make juice.
 Marinate the minced beef in the pear juice for 30 minutes.

3 Add the seasoning sauce and mix well.

4 Place a damp hanji (Traditional Korean paper) on a heated grill.
 Put the seasoned beef on it. Grill the beef, sprinkling water on the paper.

5 Put another damp piece of paper on the beef. Turn upside down and grill again.
 Sprinkle with sesame seeds and serve.

Minari jeon
(Pan-fried Water Parsleys)

Ingredients

Water parsleys 200g, eggs 100g, chopped beef 70g, rice powder 75g, flour 55g,
young green hot peppers 30g, red hot peppers 30g, salt 1 teaspoon, water 100ml, some cooking oil
For the seasoning chopped green onions 1/2 tablespoon, crushed garlic 1 teaspoon,
salt 1 teaspoon, sesame salt, a pinch of black pepper, some sesame oil

Cooking Directions

1. Cut the water parsleys into 20cm long pieces.
2. Slice the young green hot peppers and red hot peppers diagonally into 0.2cm pieces.
3. Break the eggs and mix well with water.
4. Sieve the flour and rice powder. Season with salt. Mix with the beaten egg.
5. Season the chopped beef and half-fry in a pan.
6. Line a pan with cooking oil. Place in the water parsleys and pour in the mixture from point 4.
7. Place the beef, young green hot peppers, and red hot peppers onto the water parsley mixture and fry them altogether.

Note

Be careful not to cook the vegetables too much. Thus, half-fry
the sea foods and beef first and then mix with the vegetables
and cook thoroughly together.

Doraji jeonggwa(Candied Bellflowers)

Ingredients

Bellflowers 300g, sugar 180g, honey 2 tablespoons, dextrose syrup 40g, a pinch of salt, water 400ml
Gardenia seed juice gardenia seeds 2, water 140ml

Cooking Directions

1 Rub the bellflowers in salt and cut up into 5cm long pieces.
2 Put some salt in some boiling water and parboil the bellflowers. Then soak them in cold water for 20 to 30 minutes to remove the bitter taste.
3 Bring the bellflowers, sugar, and water to the boil, skimming off the foam.
4 When the mixture is reduced by half, add the gardenia seed juice and then the dextrose syrup
  and braise them for a while over a low heat until most of the water disappears.
5 Finally, add honey.

Mu jeonggwa(Candied Radishes)

Ingredients

Radishes 200g, dextrose syrup 200g, salt 1/2 teaspoon, water 200ml

Cooking Directions

1 Cut up the radishes into half-moon shapes, with about a 0.5cm thickness.
2 Put the radishes into the water, add salt and then parboil them. Soak them in cold water
  and allow to cool. Place on another plate and allow to drain to remove the excess moisture.
3 Mix the dextrose syrup with water (200ml) and boil in a pot. Add the radishes from point 2 and braise them.

Yeongeun jeonggwa(Candied Lotus Roots)

Ingredients

Lotus roots 300g, sugar 180g, dextrose syrup 40g, honey 2 tablespoons, a pinch of salt, water 400ml
For the schizandra juice Schizandra 100g, water 100ml,
For the vinegar sauce vinegar 1 cup, water 400ml

Cooking Directions

1 Peel the lotus roots, cut them into pieces which are 0.5cm thick, and soak in the vinegar sauce.
2 Parboil the lotus roots in boiling water. Soak in cold water for a while and then leave to drain to remove the excess moisture.
3 Put the parboiled lotus roots, sugar, salt, and water into a pot and bring them to the boil, skimming off the foam.
4 When the liquid has reduced by the half, add the schizandra juice and braise them over a low heat.
5 Add the dextrose syrup and braise further. Add honey.

[Jeju]

As Jeju, with its high mountains, is vulnerable to drought and heavy winds and is a place where water is scarce it does not have a significant amount of agricultural products to offer and their cooking methods are very simple. Thus, completely different types of food were developed. Taste is ensured using only the main ingredients with only a small amount of seasoning. Due to a lack of proper storage technology, fishery products or vegetables are consumed raw and salted seafood and seaweeds are also often consumed.

Normally, people here enjoy cooked rice with mixed grains, doenjang guk (soybean paste soup), kimchi, salted seafood or raw doenjang (soybean paste), and raw or boiled vegetables.

Memil goguma beombeok
(Buckwheat and Sweet Potato Mush)*

Ingredients

Buckwheat 300g, sweet potatoes 630g, water, salt 1 tablespoon

Cooking Directions

1 Peel the sweet potatoes. Cut into 3cm thick pieces.
2 Pour some water in a pot. Put the sweet potatoes in a pot, add salt
and bring them to the boil.
3 When the sweet potatoes are almost cooked, sprinkle in the buckwheat flour
and keep stirring until they are thoroughly cooked.
4 When the buckwheat is cooked and the color becomes transparent, turn off the heat.

Note

If the sweet potatoes are produced in Jeju, the taste is even
better.

* Sweet potatoes are catching on among US women for their high efficacy for dieting.

Mojaban guk
(Gulfweed Soup)

Ingredients

Gulfweeds 220g, pork backbones 600g, radishes 100g, onions 100g, Chinese cabbage 100g, buckwheat 3 tablespoons, water 2L, young green hot peppers 30g, red hot peppers 30g, small green onions 20g, green onions 35g, red hot pepper powder 1 tablespoon, crushed garlic, crushed ginger, a pinch of salt

Cooking Directions

1. Put the pork backbones in a pot and pour water (5 cups) over them.
   Bring them thoroughly to the boil. Take out the bones and place in a bowl.
2. Trim the gulfweeds and wash them well. Cut them into 2cm long pieces.
   Cut the Chinese cabbage and small green onions into the same size pieces.
3. Cut up the radishes into rectangular pieces (2.5×2.5×0.5cm)
   and slice the onions into thin pieces (5×0.3×0.3cm).
   Chop the young green hot peppers and red hot peppers (0.3cm)
4. Combine the buckwheat and water (5 cups) and mix well.
5. Put the ingredients from point 2 and 3 into the broth from point 1.
   Bring them to the boil. Add the green onions, crushed garlic, crushed ginger and salt.
   When they begin to boil once again, add the buckwheat water from point 4.
6. Serve with red hot pepper powder.

Note

This was a dish that was served to the guests at the house hosting a party or at a mourner's house. It was made by putting in gulfweed, Chinese cabbage, buckwheat powder and pork's intestines after killing a pig the previous day, mixing them together in an iron pot and boiling them all together. This is also called 'Molmangguk', and it is consumed with kimchi, cut according to the individuals' taste, or with pepper powder, or pepper. This soup tastes the best when served hot.

Ureok kong jorim
(Braised Rockfish and Soybean)

Ingredients

Rockfish 3, soy beans 70g, young green hot peppers 15g, red hot peppers 15g
For the seasoning sauce water 4 tablespoons, light soy sauce 4 tablespoons,
red hot pepper powder 2 teaspoons, sugar 1 tablespoon, crushed garlic 1 tablespoon,
cooking oil, a pinch of sesame seeds

Cooking Directions

1 Slightly wash the soy beans in water. Stir-fry them in a pot.
2 Remove the innards of the rockfish, wash them well. Make two slits in them.
3 Chop the young green hot peppers and red hot peppers (0.3 cm).
Make the seasoning sauce with the ingredients.
4 Put the rockfish, stir-fried soy beans, and hot peppers in a pot and pour the seasoning sauce
over them. Braise them until the broth is reduced by half.

Note

Rockfish are best enjoyed from spring to summer. Black rockfish
taste much better than the normal ones.

Bing tteok
(Buckwheat Wraps)*

Ingredients

buckwheat 5 cups, water 1.6L, radishes 800g, small green onions 100g, salt 1 teaspoon, sesame salt 1 teaspoon, sesame oil 2 teaspoons, cooking oil

Cooking Directions

1 Knead buckwheat flour in tepid salty water
2 Cut the radishes into pieces (5×0.3×0.3cm) and bring them to the boil.
  Then wring out the water. Chop the small green onions into 0.3cm pieces.
3 Add the sesame oil, salt, and sesame salt into point 2 and mix well to make a filling.
4 Line a pan with oil. Scoop out a lump of the buckwheat dough and pan-fry it (20cm diameter).
5 Place the pan-fried rice cake from point 4 on a plate. Spoon out a lump of the filling from point 3 and place it into the middle of the rice cake. Roll it up and press both ends with your fingers.

Note

In the past, women in Jeju Island used to take one basket of bing tteok to pay homage to the houses carrying out religious ceremonies. Sometimes, red beans are used instead of radishes to fill up the inside. Buckwheat dough is also made thick and steamed in the shape of dumplings, which is called Memil mandu tteok. When buckwheat powder is ground after malt or soybean powder is ground in the same grinder, it is difficult to make good bing tteok since the powder becomes loose.

* It is ideal for a party dish.

Sirome cha
(Crowberry Tea)

Ingredients

Crowberries 2kg, sugar (honey) 1.5kg, water

Cooking Directions

1 Slightly wash the crowberries and remove the excess moisture.
2 Spread the sugar and crowberries alternately in a glass bottle and allow them to saturate for a month
  to turn into the crude liquid crowberries
3 Add the liquid into cold water or hot water according to individual's taste.

Note

Crowberries are native to Mt. Halla in Jeju and tastes savory.
The color is darker than black schizandra tea.

한국전통향토음식(영어)

초판 1쇄 인쇄 2020년 06월 15일
초판 1쇄 발행 2020년 06월 25일
지은이   국립농업과학원
펴낸이   이범만
발행처   **21세기사**
등록   제406-00015호
주소   경기도 파주시 산남로 72-16 (10882)
전화 031)942-7861 팩스 031)942-7864
홈페이지   www.21cbook.co.kr
e-mail   21cbook@naver.com
ISBN 978-89-8468-873-5

정가 20,000원